D0884271

RFID in Manufacturing

Oliver Günther • Wolfhard Kletti
Uwe Kubach

RFID in Manufacturing

Oliver Günther
Humboldt-Universität zu Berlin
Wirtschaftswissenschaftliche Fakultät
Inst. f. Wirtschaftsinformatik
Spandauer Str. 1
10178 Berlin
Germany
guenther@wiwi.hu-berlin.de

Uwe Kubach
SAP Research
CEC Dresden
Chemnitzer Str. 48
01187 Dresden
Germany
uwe.kubach@sap.com

Wolfhard Kletti
MPDV Mikrolab GmbH
Römerring 1
74821 Mosbach
Germany
w.kletti@mpdv.de

ISBN 978-3-540-76453-3 e-ISBN 978-3-540-76454-0

DOI 10.1007/978-3-540-76454-0

ACM Computing Classification (1998): J.1, C.3, K.6.1, K.4.3, H.2.8

Library of Congress Control Number: 2007940032

Typesetting and production: LE-TEX Jelonek, Schmidt & Vöckler GbR, Leipzig, Germany
Cover design: KünkelLopka, Heidelberg, Germany

Printed on acid-free paper

9 8 7 6 5 4 3 2 1

springer.com

Preface

Radio frequency identification (RFID) is likely to join the ranks of those information technologies that are called "disruptive." Its adoption by an enterprise and subsequent integration into the local information technology (IT) infrastructure will trigger considerable changes to existing architectures and business processes. The cost of the subsequent reengineering tasks may well exceed the cost of the required hardware and software. As a result, RFID introduction hardly comes cheap, nor is it likely to remain an isolated phenomenon.

On the other hand, RFID and related sensor technologies have the potential to change the way we control business processes in a fundamental manner. RFID allows us to track objects throughout their production and subsequent life cycle, spanning enterprise boundaries as well as spatial and temporal limits. A consequent application of the technology leads to a detailed and accurate "digital shadow" of the objects and processes being surveyed. With appropriate aggregation and reporting techniques, this information can be used by decision makers at different levels of the organizational hierarchy. This may lead to considerable operational and strategic benefits. Prototypical installations confirm this positive outlook; some of them have already led to impressive productivity gains throughout the various functional areas of an enterprise.

We believe RFID is likely to have a significant impact on a broad variety of business processes, in particular—and maybe most prominently in—manufacturing and logistics. In this book, we present some first insights into how the technology can be applied in manufacturing. We also offer operational and strategic guidelines that organizations can use to introduce the technology so that a positive return on investment is most likely.

Our practical insights are based on six case studies and involve analysis of the concrete benefits of RFID technology in manufacturing and supply chain management. The companies investigated are small and midsize manufacturers, mostly automotive suppliers but also companies from the electronics and packaging industries. They are all headquartered in Germany, but most of them have considerable export activities. Their size ranges between several hundred and 18,000 employees. The majority of the companies surveyed use SAP R/3 or some other kind of enterprise software [also called an enterprise resource planning (ERP) system]. Some also use a manufacturing execution system (MES) such as MPDV's HYDRA.

All of the companies surveyed see considerable potential for RFID in manufacturing. RFID is expected to lead to increased automation, especially in data capture, and therefore to reduced labor costs. Improved tracking and tracing may lead to a more stable manufacturing process, with interruptions in the production process becoming less frequent. This should help reduce downtime, lower error rates, and cut down on production waste. Tracing faulty parts and processes in the wake of a complaint or an accident will become much easier; given the increasing demands regarding product liability, this is likely to create major competitive advantages for early adopters. In container management, RFID can optimize scheduling and help reduce shrinkage. Using RFID for the uniform labeling of shipments may lead to considerable savings in labor and hardware.

In order for these positive potentials to come true, it is crucial that RFID not form a technology island but be tightly integrated into existing IT infrastructures. ERP and MES systems need to be prepared to take advantage of the rich data becoming available through RFID. Appropriate filtering techniques need to be put into place to make sure that ERP and MES components receive the relevant information in the appropriate granularity. Moreover, companies must carefully consider how to distribute storage and processing in the resulting multitier IT architecture that ranges from RFID tags and sensors to data warehouses and business intelligence tools.

When developing our case studies, we found that most of today's RFID applications focus on issues that are *operational* and *local*—that is, within the enterprise. In many cases, this is most likely to guarantee a short-term return on the required investment. Cases in which RFID is used as a *strategic* enabler, on the other hand, are found much less frequently. The same holds for *inter-enterprise* applications, in which supply chain partners cooperate to maximize the positive impact of the new technology. This may be done, for example, by leaving RFID tags on the objects being produced as they move through the supply chain and by integrating the related business processes. Cooperating partners can use the technology to provide fine-grained product traceability and quality assurance across the whole supply chain. This may translate into significant and tangible competitive advantages.

The main reason such an RFID-enabled supply chain integration is not being considered by a larger number of enterprises is that the specific costs and benefits are not always correlated. Some companies in the supply chain may incur considerable costs that outweigh the local benefits, and vice versa. This can lead to a classical prisoner's dilemma: It could well be possible that an existing supply chain could gain considerably from introducing RFID technology. These gains, however, are never realized because some participants would need to incur costs that are not justifiable in comparison to their own benefits. As a result, they decide—for completely rational reasons—not to adopt the new technology. One way to break this deadlock is to negotiate compensation payments between different participants in the supply chain with the objective of distributing the benefits fairly among the participants, a model that we will discuss in greater detail later in this book.

The book is structured as follows: Chapter 1 is an extended introduction, presenting the relevant standards and outlining the essential productivity potentials of RFID. In Chaps. 2 and 3 we present the salient features of enterprise software (ERP systems) and manufacturing execution systems. Chapter 4 continues with a detailed presentation of the six case studies described above. Chapter 5 discusses the lessons learned from these case studies and presents a number of operational and strategic recommendations. Chapter 6 concludes with a summary and outlook.

We would like to thank all those colleagues, coworkers, and graduate students who contributed to this book directly or indirectly. Lenka Ivantysynova and Holger Ziekow of Humboldt-Universität conducted the case studies and coauthored Chaps. 1, 4, and 5. Christof Bornhövd, Gregor Hackenbroich, Stephan Haller, Tao Lin, Jochen Rode, and Joachim Schaper of SAP contributed to Chap. 2. Jochen Schumacher of MPDV established many of the corporate contacts and contributed to the case studies. Franziska Brecht, Derya Saki, Mert Sengüner and Matthias Schmidt, all of Humboldt-Universität, helped us with the case studies, the figures, the index, and various other editing tasks. Finally, we want to thank Ralf Gerstner of Springer-Verlag for his many suggestions and for his patience during the "manufacturing" of this book.

Berlin/Heidelberg/Dresden
October 2007

Oliver Günther
Wolfhard Kletti
Uwe Kubach

Contents

List of Contributors

Dr. Christof Bornhövd
SAP Research Center Palo Alto, LLC, 3475 Deer Creek Road, Palo Alto, CA 94304, USA

Prof. Oliver Günther, Ph.D.
Institute of Information Systems, Humboldt-Universität zu Berlin, Spandauer Str. 1, 10178 Berlin, Germany

Dr. Gregor Hackenbroich
SAP Research CEC Dresden, Chemnitzer Straße 48, 01187 Dresden,Germany

Stephan Haller
SAP Research, CEC Karlsruhe, Vincenz-Priessnitz-Strasse 1, 76131 Karlsruhe, Germany

Lenka Ivantysynova
Institute of Information Systems, Humboldt-Universität zu Berlin, Spandauer Str. 1, 10178 Berlin, Germany

Wolfhard Kletti
MPDV Mikrolab GmbH, Römerring 1, 74821 Mosbach, Germany

Dr. Uwe Kubach
SAP Research CEC Dresden, Chemnitzer Straße 48, 01187 Dresden, Germany

Nadine Lenz
MPDV Mikrolab GmbH, Römerring 1, 74821 Mosbach, Germany

Tao Lin, Ph.D.
SAP Research Center Palo Alto, LLC , 3475 Deer Creek Road, Palo Alto, CA 94304, USA

Dr. Jochen Rode
SAP Research CEC Dresden, Chemnitzer Straße 48, 01187 Dresden, Germany

Dr. Joachim Schaper
SAP Research Center Palo Alto, LLC, 3475 Deer Creek Road, Palo Alto, CA 94304, USA

Jochen Schumacher
MPDV Mikrolab GmbH, Römerring 1, 74821 Mosbach, Germany

Holger Ziekow
Institute of Information Systems, Humboldt-Universität zu Berlin, Spandauer Str. 1, 10178 Berlin, Germany

List of Abbreviations

AIR	Manufacturer of airbags
ALE	application level events
API	Application programming interface
ASN	Advanced shipping notification
ATP	Available-to-promise
B2B	Business-to-business
B2MML	Business-to-manufacturing markup language
BPP	Business process platform
CAS	Manufacturer of cast parts
CEP	Complex event processing
CIM	Computer integrated manufacturing
CLU	Manufacturer of sliding clutches
COO	Manufacturer of engine cooling modules
COM	Component object model
CON	Manufacturer of connectors
DC	Device controllers
DCOM	Distributed component object model
DCS	Distributed control system
DNS	Domain name service
ECP	electronic product code
EDI	electronic data interchange
EPCIS	EPC information services
ERP	Enterprise resource planning
FDA	U.S. food and drug administration
GPS	Global positioning system
HF	high frequency
HMI	Human–machine interface
ID	Identification
IP	internet protokoll
JIS	just-in-sequence
LED	light-emitting diode
LF	low frequency
MES	Manufacturing execution system
MM	Materials management
OEE	overall equipment efficiency
OEM	Original equipment manufacturer

OLE	Object linking and embedding
ONS	Object name service
OPC	OLE for production control
OPC-UA	OLE for production control unified architecture
P2B	Plant-to-business
PAC	Manufacturer of packaging
PDA	personal digital assistant
PDC	Production data collection
PLC	Programmable logic controller
PLM	Physical markup language
RFID	Radio frequency identification
ROI	Return on investment
SCADA	Supervisory control and data acquisition
SOA	Service-oriented architecture
TCO	Total cost of ownership
UHF	ultrahigh frequency
VDI	Verein Deutscher Ingenieure
WIP	Work in process
XML	Extensible markup language

Chapter 1
RFID in Manufacturing: From Shop Floor to Top Floor

Coauthored by Lenka Ivantysynova and Holger Ziekow

"As manufacturers gain control over inventory and grab accurate demand signals from their customers, the last blind spot in the global supply chain is the plant floor."
Rob Spiegel, Automation World

1.1 Architectural Perspectives

The vision of an enterprise information system comprising a complete virtual image of the enterprise is as old as information technology (IT) itself. IT architectural diagrams from the 1960s often take a pyramidal shape, thus capturing the hierarchical structure of most enterprises. Staff members were at the bottom of this pyramid, entering data into the IT system (Fig. 1.1). After undergoing various phases of filtering and aggregation, the data were compiled into a variety of tactical and strategic reports for decision makers positioned at higher levels of the hierarchy.

When this vision first appeared, several factors prevented it from being a great success. Data entry was usually manual and therefore costly, incomplete,

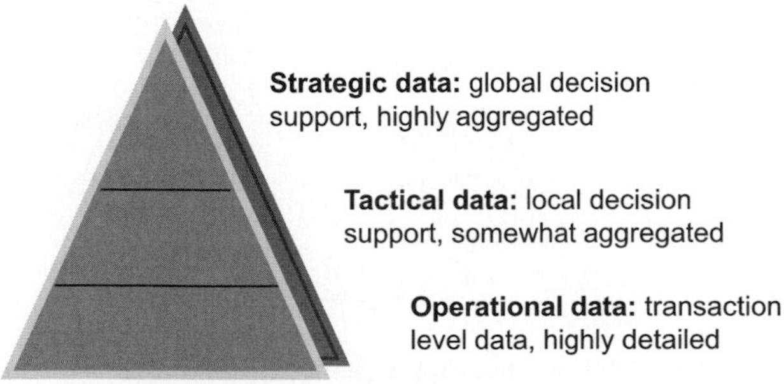

Strategic data: global decision support, highly aggregated

Tactical data: local decision support, somewhat aggregated

Operational data: transaction level data, highly detailed

Fig. 1.1 Simple management information system pyramid architecture

O. Günther, W. Kletti, U. Kubach, *RFID in Manufacturing*
DOI: 10.1007/978-3-540-76454-0, © Springer 2008

and error prone. Filtering and aggregation algorithms were much less powerful than the technologies in use today. As a result, the quality of the reports delivered to senior staff was often questionable. Moreover, the only way to access this information was through printed reports. User-friendly online access via a PC was a distant possibility at best. Executives thus quickly learned not to rely on their corporate IT, and many preferred to follow traditional ways of decision making.

IT progress of the last decades, including the development of radio frequency identification (RFID) and other sensor technologies as an important component of enterprise computing, has led to a situation that is fundamentally different from the situation described above. This progress has already led to impressive productivity gains throughout the various functional areas of enterprises. Atkinson and McKay (2007) give a thorough overview of how IT has transformed our economies and led to tangible economic benefits. Niederman et al. (2007) present the potential of RFID in supply chain management. They use a data life-cycle framework to discuss the key issues involved when introducing RFID into an existing IT infrastructure.

Today, data capture has been automated along large parts of the supply chain. Algorithms for aggregation, search, and data mining have improved tremendously, both in terms of functionality and performance. The vision of a “digital shadow” or “virtual image” for each physical object seems closer than ever. This creates major privacy challenges and raises difficult questions concerning an integrated enterprise IT architecture. But it also opens up new perspectives for event and object tracking inside a corporation and for inter-enterprise cooperation, spanning the complete product life cycle starting with production, continuing through distribution, and finishing with aftersales services and possibly recycling.

Figure 1.2 illustrates the kind of information flow along the supply chain that is already possible. In the example pictured in the figure, only the objects

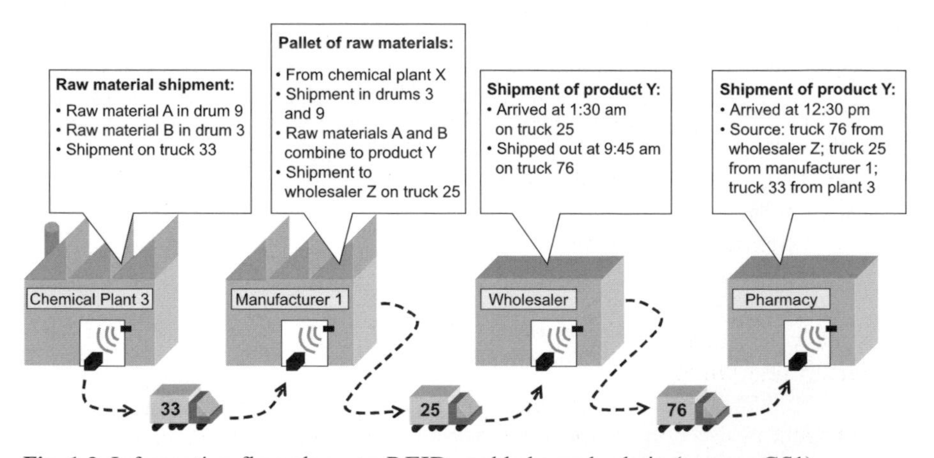

Fig. 1.2 Information flow along an RFID-enabled supply chain (source: GS1)

being produced are being tagged. Tagging could be extended to production machinery and tools, thus allowing even tighter control of the production process and improving inventory control. Even personnel could be tagged. Given the obvious concerns regarding privacy and loss of control, however, any potential benefits need to be carefully gauged against the considerable drawbacks.

The term "real-world awareness" is increasingly being used to characterize this convergence between physical and virtual worlds and the availability of timely and accurate information. Obvious advantages include the following:

- Optimized inbound and outbound logistics via more accurate visibility to production activities
- Efficiency gains by giving plant personnel access to up-to-date location information for materials, assets, and orders
- Increased responsiveness to unplanned events

In this book we present the results of a variety of case studies analyzing the benefits of RFID in manufacturing and supply chain management. Industries surveyed include automotive suppliers as well as companies from the electronics and packaging industries. The majority of the companies surveyed use SAP R/3 or some other kind of enterprise software [also called an enterprise resource planning (ERP) system]. Some are also using a manufacturing execution system (MES) such as MPDV's HYDRA. If RFID and sensor technologies are added to such an architecture, various integration issues need to be addressed:

- Which of the RFID and sensor data need to be forwarded to the MES layer?
- Which of the RFID and sensor data need to be forwarded to the ERP layer?
- What kind of filtering needs to be applied and when?
- What kind of information should be exchanged with which partners in the supply chain?
- What kind of sensor technology is most cost efficient in a given situation? What about barcodes?

RFID applications are normally closely tied to the MES controlling the production process. The typical functionalities of an MES are described by the Manufacturing Enterprise Solutions Association:

- Operations scheduling and production control
- Labor management
- Maintenance management
- Document control
- Data collection
- Quality management
- Performance analysis

RFID technology may support most of these functionalities. In operations scheduling and production control, RFID can be used for guaranteeing process safety and interlocking. If materials or material containers are equipped with a unique ID (provided via barcode or RFID), the MES can ensure that all preceding process steps have been conducted successfully before starting the next manufacturing step. Furthermore, production order data and manufacturing parameters may be written to the RFID tag at the first manufacturing step and then read and updated locally, providing fast local data maintenance and redundancy for the MES.

Concerning labor management, plant personnel could automatically be registered via appropriate sensor technologies. Also, location tracking of plant personnel could potentially be enabled. Once again, privacy concerns would need to be taken into account when considering such measures. In maintenance management, data may be stored locally at the resource in an RFID tag. This may reduce the required paperwork for performing maintenance and updating associated records. Regarding data collection, RFID can help automate the tracking and tracing of materials, work in process (WIP), the location of mobile resources, etc. For quality management, data about the quality targets may be stored locally at the material within an RFID tag.

Sensor gates and RFID readers collect data from the shop floor. They pass this data on to "edge servers." This term refers to computers that are directly connected to the RFID readers and sensor gates. Edge servers belong to the back-end system. The back-end system comprises the edge server software, the ERP system, and the MES. Figure 1.3 visualizes the basic functionalities of the resulting three-tier architecture integrating sensor data into an existing ERP/ MES installation. It also shows the possibility of opening up an IT architecture to customers, suppliers, and other business partners in the supply chain.

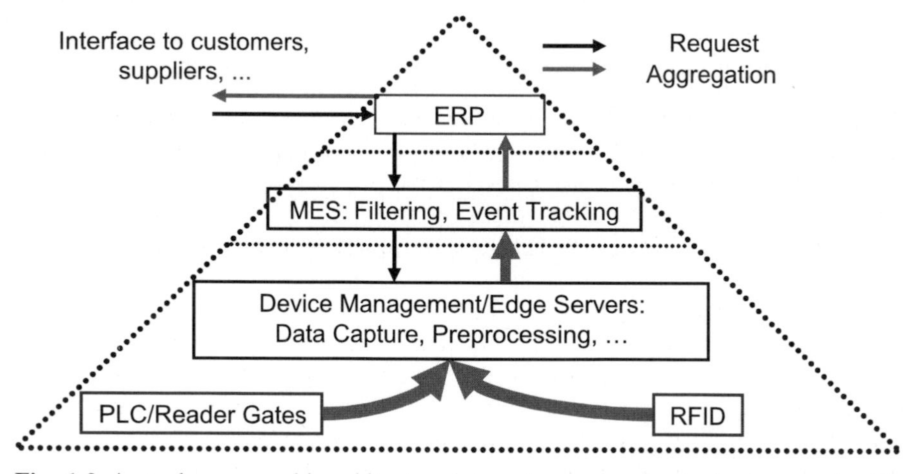

Fig. 1.3 A modern pyramid architecture integrating sensor/radio frequency identification data, a manufacturing execution system, and an enterprise resource planning system

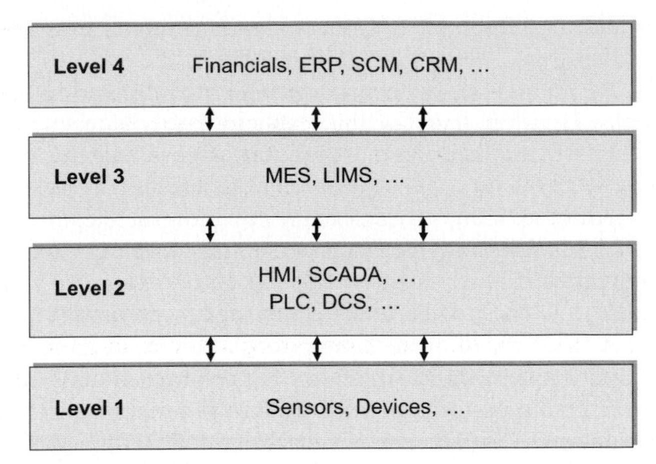

Fig. 1.4 ISA-95 functional reference model based on the Purdue reference model, showing connectivity from shop floor to top floor for adaptive manufacturing

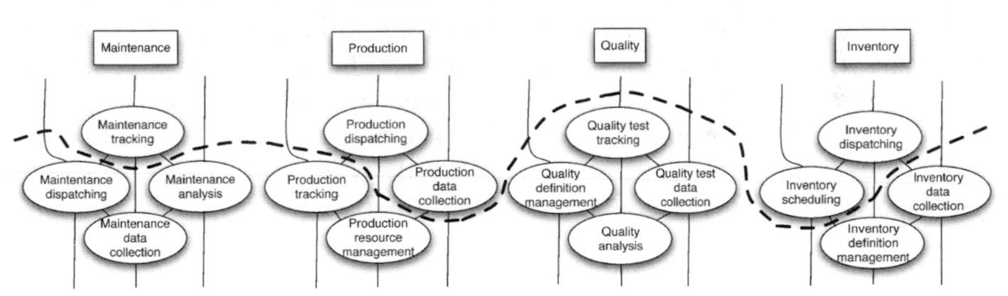

Fig. 1.5 Fuzzy borders between different tiers of the architecture (source: ISA S-95 standard)

A more detailed distribution of the different functionalities has been proposed in the ISA-95 functional reference model (Fig. 1.4), but this should not convey a false impression of accuracy. For many functionalities, one has several choices regarding how to assign them to tiers. The border between hardware controllers and MES is equally fuzzy, as is the border between MES and ERP (Fig. 1.5).

Moreover, some enterprises prefer not to use an MES, connecting the hardware controllers directly to the ERP system. This is only possible, of course, if the ERP system is equipped to do this by containing modules that accept the controller output as input and that perform the necessary modeling and filtering. The SAP Auto-ID Infrastructure (see Chap. 2) provides such functionality. Moreover, vendors such as Infosys (Deshpande and Singh 2006) and PEAK (Schultz 2007) offer middleware components that serve as interfaces between the hardware on the shop floor and the ERP software. Often, however, this includes only a technical interface without any application or process support, as is provided, for example, by an MES (Kletti 2006).

Currently, RFID and sensor technologies are mainly used to automate logistics processes. However, there are many other ways these technologies can be employed to make manufacturing more efficient. To evaluate the feasibility and cost efficiency of an RFID solution, one needs to address the following issues:

- Continuous or other inline processes may not allow the tagging of "discrete" entities and thus will dramatically increase the complexity of the tracking and tracing problem.
- High-volume, low-cost goods rarely justify the cost of unique identifiers.
- The small physical size of critical components cannot support a tag.
- Complex and frequent transformations of the physical goods may make it difficult to maintain up-to-date identity.
- The combination of RFID and sensor technology within the walls of the manufacturing plant or distribution center often requires substantially higher granularity than tracking unit loads across the supply chain.
- In the context of combined RFID and location sensing, a substantial amount of time and resources may be required to locate and expedite materials and orders within a location. The combination of RFID and global positioning system (GPS) location tracking is not typically suitable or accurate enough for in-plant tracking. Other "smart-tag" technologies, which may include three-dimensional position tracking, may be required.
- Concerning deployment strategies, it may be preferable from a cost and operational perspective to tag carrier devices (carriers, pallets, bins) instead of individual items and to maintain a dynamic association of specific items and orders with these devices.

Finally, one word about terminology: Note that the terms "manufacturing" and "production" are often used interchangeably in the literature, even though the term production typically signifies a broader range of activities than the term manufacturing does. Wikipedia defines manufacturing as "processing raw materials into finished goods" and regards it as a "specific use" of production. Production, on the other hand, is defined in a microeconomic sense as "the act of making things." This involves "decisions…on what goods to produce, how to produce them, the costs of producing them, and optimizing the mix of resource inputs used in their production." This latter definition goes somewhat beyond what is commonly meant by the term manufacturing. Nevertheless, for this book we opted for better readability and decided to use the two terms synonymously, unless noted specifically.

1.2 RFID Basics and Standards

For our discussion of the various standards, we use a reference RFID environment consisting of four elements:

- *RFID tags:* The tags (also called RFID chips) are attached to physical objects and store at least a unique identifier of the object that they are attached to. In addition, they might store some other user data.
- *RFID readers:* These are the hardware devices that directly interact with the RFID tags. Higher-level applications can access RFID readers through a well-defined protocol. RFID readers provide at least reading and in some cases also writing functionality. In addition, they might offer functionalities for aggregating or filtering read operations and for disabling RFID tags either temporarily (e.g., via locking) or permanently ("kill").
- *RFID middleware:* The middleware is software that can run centrally on a single server or be distributed over different machines. Its major role is to coordinate a number of RFID readers that are usually located close to each other, for example within a single plant or production line. The middleware buffers, aggregates, and filters data coming in from the readers to reduce the load for the applications.
- *Applications:* The RFID data may be used by a great variety of software systems. In a manufacturing environment, the applications are typically part of the MES or ERP system.

After introducing some technical foundations, we present three families of standards that cover various aspects of an RFID environment. The only standards that have been designed especially for RFID environments are the *EPCglobal*™ standards. These standards cover all aspects from tag to application. The *OPC/OPC-UA* standards family primarily addresses communication with systems close to the shop floor, such as programmable logic controllers (PLCs) or plant historians. In an RFID scenario, OPC/OPC-UA would be most suitable for the communication between the middleware and the readers. As of today, OPC is not broadly accepted for this purpose. However, it is an important standard for shop floor integration with IT systems in general and is used when RFID readers and shop floor equipment must be used in combination. The third standards family that we consider is the *ISA S-95* family, for plant-to-business integration with a special focus on manufacturing.

1.2.1 Technical Foundations

Currently, the following recognition systems are used in manufacturing:

- *Visual recognition:* Different colors and shapes are used for identification.
- *Contact recognition:* Systems with, for example, reed relay belong to this category. Such systems have the disadvantage of mechanical wear and low identification depth, in most cases between only 6 and 8 bits.
- *Optical recognition:* Barcode and other visual systems have the disadvantage of depending on a line of sight between object and reader. (See Fig. 1.6.) Dirt, vapor, refraction, scratches, and even vibrations can interfere with the automatic recognition. In particular, the traditional one-dimensional bar-

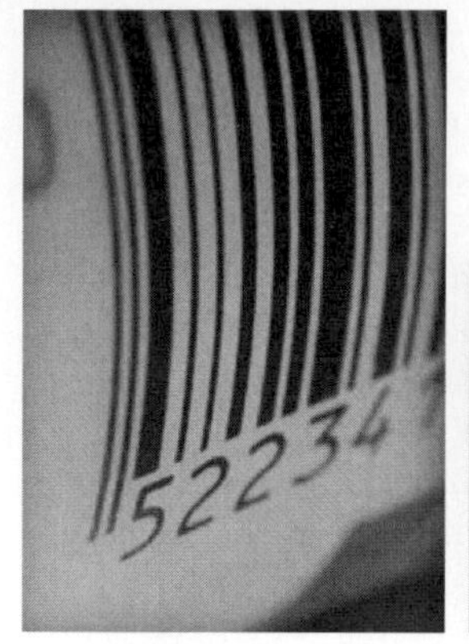

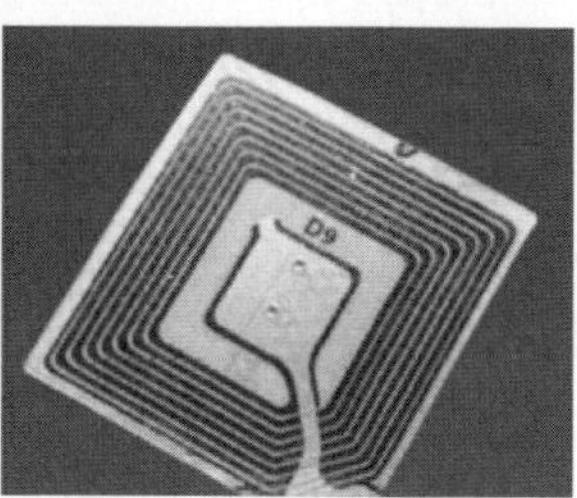

Fig. 1.6 Example of a barcode label and an RFID chip

codes are rather sensitive to disturbances. Two-dimensional barcodes are able to store much more information but, in turn, require more space. Matrix codes mitigate this problem somewhat, but they require a more sophisticated logic in the evaluating software.

- *Radio frequency identification:* RFID solves many of the problems discussed above. In particular, it works over a certain distance and even through non-metallic materials. We will now discuss RFID in more detail.

An antenna emits radio waves generating voltage in the inductor of the passive transponder or triggering the active transponder to send data (Fig. 1.7). The transponder chip starts working with this voltage, uses the inductor as antenna, and sends its ID to the reader antenna in bit-serial form (00101111100…). The transponder signal is evaluated in the decoder, checked for errors, converted into a code (000F5A3B1C…), and forwarded for further processing.

Low frequency (LF; 125 kHz): Low-frequency systems have proven sufficient in many contexts even though their range is limited to 1–2 m. Successful implementations exist in manufacturing, assembly, logistics, and access control. The transponders are not very expensive and work even when embedded in metal. However, transponders are not protected against collision; bulk reading of many transponders at once is not possible.

High frequency (HF; 13.56 MHz): This higher frequency enables thin and inexpensive write/read transponders with anticollision technology, thus enabling the reading of several transponders at the same time. This technology represents a good compromise between cost and benefit even though it is still too expensive to lead to the "disposable transponder." Moreover, the transponders are not very resistant to adverse mechanical and thermal conditions, and they do not work very well in environments with lots of metal.

Active ultrahigh frequency (UHF; 868 MHz): This technology based on ultrahigh frequencies allows read distances of up to 100 m at high transport

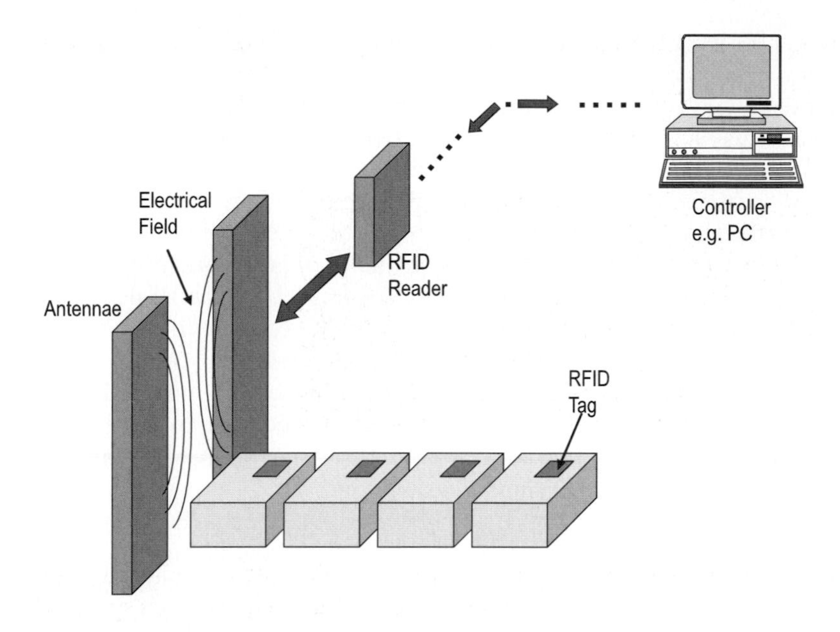

Fig. 1.7 A typical architecture for RFID data capture

rates, provided directional radiation or an antenna field. As these data carriers dispose of a separate energy source, multifunctional transponders are possible, registering via light-emitting diode (LED) or disposing of an integrated temperature logger. In this case, multiple transponders can also be read simultaneously via anticollision technologies that can manage up to 2,000 tags in the reading area. Yet the high costs for the transponders and readers as well as the limitations of the temperature range due to the battery must be considered disadvantages. Metal reflections may also prevent the position from being assigned uniquely.

Passive UHF (868 MHz): Passive label transponders using 868 MHz technology also have the advantage of combining write/read transponders with anticollision technology at an attractive cost–benefit ratio. They are especially well suited for label and track pallets and cartons. By means of directional radiation, ranges of up to 6 m can be reached at high transport rates. However, these transponders are typically not very robust with respect to adverse mechanical and thermal conditions. Location problems in the presence of metal as well as the relatively high costs for readers constitute other disadvantages.

In recent years the barcode has been established in many areas, including manufacturing. To identify goods or objects within manufacturing, this code carrier has become absolutely indispensable. Careful analysis is therefore required to see where an existing barcode solution should be replaced by RFID or whether a combination of barcode and RFID may be advisable. This is often a cost–benefit issue. RFID solutions are mostly more expensive, but they may well be worth the price. Sometimes, however, the benefits of an RFID solution do not occur at the same site where the costs occur. In a supply chain, the investments of one participant may benefit other participants, and special measures may have to be taken to assure fairness and avoid free-riding.

In general, one has to consider the total system costs, not just the price of the code carriers. If only a few code carriers but many readers are needed, the price of the readers and IT infrastructure tip the scales. If, on the other hand, in a high-rack warehouse thousands of spare parts or goods and only a few goods movements are to be covered, the price of the single code carrier determines the expenses. The basic technical advantage of RFID compared with barcodes or other code carriers is that the code carrier may store many pieces of information on very little space and that this information can be modified, extended, or exchanged automatically, without requiring any contact and without having to exchange the code carrier. Moreover, there are environmental conditions in which the benefits of RFID chips clearly prevail.

Over the past years, processes associated with creating, changing, applying, and removing labels have been playing a decisive role in logistic production processes. Relevant applications include material tracking, the identification of WIP materials, random storage, and the identification of products and their use (such as in the automotive sector for the delivery of spare parts to automotive manufacturers). This demand involves the increasing transition to flexible production processes and also the individualization of products as described beforehand.

This transition is accelerated by decreasing lot sizes, smaller production and packing units and thus an increasing number of operations. Because nearly all pieces of information saved on a paper or plastic label can be put on an RFID tag, this will become an interesting application in the future to reduce manual interventions as well as the number of needed label printers, materials (labels), and administration of this infrastructure (for example, printer maintenance and networking). Certainly, similar equipment and IT infrastructure are also required for RFID chips, namely the use of write/read devices to add information or to read it on RFID tags. However, it is, of course, much more cost efficient to write RFID tags than to administer label printers.

A major disadvantage of RFID for the identification of products is that, along with barcodes, labels also contain information in plain text that employees can read and recognize without other technical tools. If information is saved on RFID tags, only stationary or mobile readers can render them visible.

1.2.2 EPCglobal™

EPCglobal™ is a nonprofit organization founded in 2003 by the global standardization consortium GS1. Its goal is to define global standards for the use of RFID technology. Like GS1, EPCglobal™ has members around the world. Its technical roots lie in research conducted by the Auto-ID Center, a global research network driven by the Massachusetts Institute of Technology with locations in Adelaide, Cambridge, Fudan, Keio, Shanghai, and St. Gallen. In 2003 the Auto-ID Center handed over its administrative functions to EPCglobal™ and continued as a pure research organization, now known as the Auto-ID Labs.

EPCglobal™ envisions an "Internet of Things" as an information network, also called the EPCglobal™ Network, that will allow producers, retailers, and consumers to easily access and exchange information about products. The information access is based on the Electronic Product Code (EPC), a globally unique identifier that unambiguously identifies each product and can be used as a reference to information stored in the EPCglobal™ Network. In contrast to barcodes, the EPC identifies each product instance rather than only product categories. To ensure the uniqueness of each product code, only so-called EPC managers are allowed to assign product codes to physical objects, and they can assign only codes from code blocks that are under their responsibility. The code blocks are assigned to the EPC managers by an issuing agency.

EPCglobal™ has defined and is still in the process of defining a number of standards that come together under the umbrella of the EPCglobal™ Architecture Framework. This architecture describes how the different components and standards suggested by EPCglobal™ will fit together to form the EPCglobal Network. Two key components of this architecture are the Object Name Service (ONS) and the EPC Information Services (EPCIS).

The EPCIS store the information that can be referenced through the product codes and is potentially exchanged between trading partners. These services are run by EPCglobal™ subscribers or EPC managers. The stored information includes static data about the products that does not change over a product's life cycle, e.g., expiration date, as well as transactional data, e.g., business transactions or product sightings.

The ONS is a look-up service that maps an EPC to the address of one or more EPCIS that hold information about the EPC. Each EPC manager runs a local ONS service that holds, for each EPC under his or her management, the address of an EPCIS service holding information for that EPC. Moreover, EPCglobal™ controls a Root ONS service. To look up an EPCIS for a specific EPC, an EPCglobal™ subscriber first addresses this Root ONS service. This service returns the address of the local ONS of the EPC manager responsible for the requested EPC. Finally, the subscriber uses this address to contact the local ONS and gets the EPCIS in question as a response. To ensure scalability, the ONS is implemented as an application of the Domain Name Service (DNS) used in the World Wide Web to map domain names to Internet Protocol (IP) addresses.

Besides these decentralized services, the EPCglobal™ Architecture Framework covers the components that each EPCglobal™ subscriber needs to read EPC tag data and participate in the EPCglobal™ Network. Actually, most of the EPCglobal™ standards finalized up to today refer to these local components. Figure 1.8 gives an overview of these components and the standards specifying the communication between them.

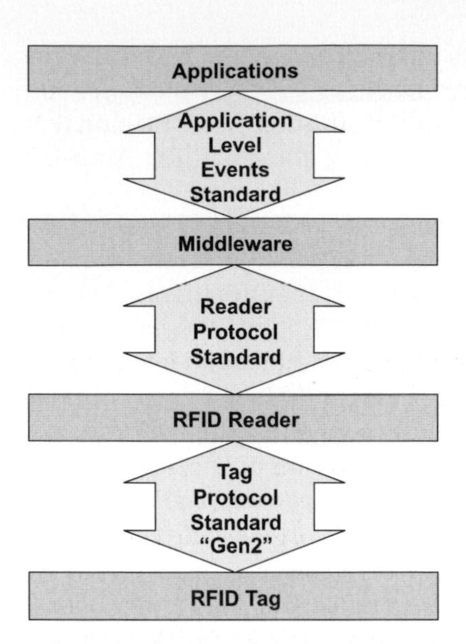

Fig. 1.8 Local components and protocols between them

The RFID tag holds the EPC and perhaps some additional user data. It might support features such as locking, access control, or killing of the tag for privacy reasons. Details are specified in the EPC Tag Data Standard, a ratified EPCglobal™ Standard. The description of this and also the following described standards can be downloaded from the EPCglobal 2005 Web site (EPCglobal 2005).

The physical communication between a reader (sometimes called the interrogator) and the tags is specified in another EPCglobal standard, the Gen 2 standard (EPCglobal 2005). Besides sending commands from a reader to a tag and receiving the answer from the tag, this standard addresses how to deal with multiple tags within a reader's range and how to minimize interference between different readers and tags.

The Reader Protocol Standard (EPCglobal 2005) describes how software applications including the middleware interact with the RFID reader. It provides a standardized way for host applications to ask RFID readers for the EPC codes in their reading range or for the additional user data on the tags. It also describes how to use additional functionality provided by the readers, such as writing to tags.

The Application Level Events (ALE) standard (EPCglobal 2005) describes how client applications can access the middleware in order to read EPC data from various sources. The ALE interface allows client applications to be completely agnostic of the reader infrastructure (e.g., the number of readers or their make and model). In addition, it provides a means for client applications to specify what to do with EPC data, such as how to filter or aggregate it.

Although the EPCglobal™ standards are so far not broadly accepted within the manufacturing domain, there are strong indications that this might change in the near future. A number of RFID tag and reader producers support the EPCglobal™ standards, and application vendors also have first products that support the EPCglobal™ standards—SAP with its Auto-ID Infrastructure, for example.

1.2.3 OPC and OPC-UA

Like EPCglobal™, OPC is also a family of different standards. The goal is to ensure the interoperability of different shop floor devices with each other as well as to provide a standardized way for applications to communicate with shop floor devices including sensors, PLCs, and historians. The OPC Foundation (OPC Foundation 2007) is the organization responsible for the OPC standards. It offers free tools for its members to test OPC compliance. Initiated in 1996 by a task force of a handful of companies from the automation industry, today the organization has more than 300 members. Originally, the OPC standard was based on Microsoft's Object Linking and Embedding (Distributed) Component Object Model (OLE COM/DCOM) technology, and therefore the name stood for "OLE for Production Control". But now the OPC standards are no longer restricted to COM/DCOM technology; therefore, the name has been changed to stand for "Openness, Productivity and Collaboration."

The first standard in the OPC family was the OPC Data Access Specification. It is still the most important standard within the family and rules the acquisition of data from shop floor devices through production control stations, MES, and even ERP systems. An OPC server running on a shop floor device or industrial PC close to the shop floor communicates via field buses with the data sources and exposes the data in a standardized way to higher-level applications, the OPC clients. With the new OPC XML-DA standard, a Web-service-based data access standard is provided. It allows running of OPC components on non-Microsoft systems.

The OPC Alarm & Events Specification provides a means for OPC clients to register for certain events or alarm conditions. The clients receive messages only when something happens that is of interest to them instead of having to read data streams continuously, as would be necessary if OPC Data Access were used.

OPC Batch specifies interfaces for the exchange of information about equipment capabilities and operating conditions, with a special focus on batch processes. OPC Data exchange specifies the communication between OPC servers via field buses. In contrast to the OPC Data Access standard, which deals with real-time data only, the OPC Historical Data Access standard specifies how data that is already stored in a system, such as a plant historian, can be accessed. OPC Security defines how the sensitive data on OPC servers can be protected against accidental or intentional manipulation.

OPC Unified Architecture (OPC-UA) is the latest specification from the OPC foundation. The first parts of this new specification were finalized in June

2006. OPC-UA provides uniform Web-service-based access to the various, formerly separated functionalities, including OPC Data Access, OPC Alarm & Events, and OPC Historical Data Access. It also overcomes the dependence on Microsoft COM/DCOM with the specification of its own communication stack. This provides the ability to run OPC-UA on non-Microsoft systems, including embedded systems, which are becoming more and more important as OPC servers.

Although the OPC standards are not very common so far for communication with RFID readers, they could provide a comfortable way to integrate RFID readers with existing plant IT infrastructures.

1.2.4 ISA-95

ANSI/ISA-95 (internationally standardized under ISO/IEC-62246) is the most prominent plant-to-business (P2B) integration standard relevant for manufacturing. The description of the standard is divided into five parts:

- ANSI/ISA–95.00.01 Models and Terminology, published in 2000 (Brandl 2000)
- ANSI/ISA–95.00.02 Object Model, published in 2001 (Brandl 2001)
- ANSI/ISA–95.00.03 Activity Models of Manufacturing Operations, published in 2005 (Brandl 2005)
- ANSI/ISA–95.00.04 Object Model of Activity Models from Part 3, unpublished draft
- ANSI/ISA–95.00.05 Business to Manufacturing Transactions, unpublished draft

The first part of this standard defines a terminology and concepts to structure manufacturing systems and operations. It builds on the Purdue reference model for computer-integrated manufacturing (Williams 1992) to define a multilevel functional reference model for manufacturing systems. Figure 1.4 illustrates how the functionality within a manufacturing system is distributed across different levels. Each of these comprises manufacturing functionality on a particular level of abstraction. In particular, the following areas are addressed:

- *Level 0:* Physical processes (machines); location of the actual production processes.
- *Level 1:* Sensing and manipulating of physical processes (sensors, actuators, RFID readers)
- Level 2: Monitoring and controlling of physical processes (PLC, human–machine interface)
- *Level 3:* Manufacturing operations and control: dispatching production, detailed production scheduling, reliability assurance, etc.; maintenance of records and optimization of the production process. Relevant time frame: seconds to days

- *Level 4:* Business planning and logistics: plant production scheduling, production, material use, delivery and shipping, inventory management. Relevant time frame: days to months

A complete RFID solution always spans multiple levels as defined in ISA-95. RFID tags are typically attached to material or material containers that are tracked through the process. Machine tools, inventory locations, and even workers may also be identified through RFID tags. The RFID tags themselves correspond to ISA-95 level 0. Hardware such as RFID readers offers level 1 functionality by reading from or writing information to RFID tags. Plant-local control devices such as a PLC or terminal PCs with an RFID device controller aggregate RFID read events and provide the interface to higher-level control systems such as an MES or ERP system. MESs are classified as level 3 systems and are responsible for orchestrating the manufacturing processes in the factory. This includes responsibilities such as operations scheduling, production control, and labor management. Level 4 functionality is typically provided by ERP systems such as the SAP ERP application.

The distinction of clearly separated levels constitutes a reference model but often does not correspond to reality. The functionality of certain levels may be combined into an integrated system. For example, an ERP system may include an RFID integration component, such as SAP Auto-ID Infrastructure, which enables it to communicate directly with manufacturing or logistics processes. In the future, the strict separation into levels and system boundaries is likely to disappear in favor of a more modular and flexible service-oriented architecture as described in Chap. 3.

ISA-95 addresses the interfaces between levels 3 and 4, describing the information that is communicated between the MES and the back-end ERP system. In part 2, the standard provides abstract definitions of information models, describing a production order, the equipment to be used for the execution, personnel, material, and other production-related entities. The ISA-95 standard defines these information models but does not offer an implementation or syntax. The business-to-manufacturing markup language, or B2MML, fills that void and fully implements ISA-95 as a set of XML schemas, with one schema per information model.

RFID itself is not explicitly considered within ISA-95. However, many of the ISA-95 information models include identification information such as data to identify a particular material or personnel. These data may be contained on an RFID tag attached to a material or carried by plant-floor personnel.

Other P2B integration standards are RosettaNet (RosettaNet 2007) and OAGIS (Open Applications Group 2007). In contrast to these standards, ISA-95 focuses solely on the integration of ERP and MES. Both RosettaNet and OAGIS go far beyond P2B integration in an attempt to model every class of business-to-business (B2B) transaction. The cost of this approach is the lack of depth for P2B integration. However, a recently started cooperation between ISA and OAGIS is likely to further harmonize the currently competing standards. ISA-95, although still far from being mainstream, has already been deployed successfully by a number of businesses and is currently the leading standard for P2B data exchange.

1.3 RFID Potentials

Judging from the case studies presented later in this book, many companies consider the introduction of RFID without a clearly defined business case. This is not altogether irrational. Many enterprises realize the potential of this technology and want to get a head start in turning this potential into a competitive advantage. Moreover, they want to be prepared in case their business partners or customers make corresponding demands. The situation of the retail industry in 2004/2005, when large retailers such as Wal-Mart and Metro put unexpected demands on their suppliers, serves as a warning to many smaller and midsize manufacturers in particular.

Despite the usual lack of a clear business case, the companies we surveyed focus on a few application areas where RFID and sensor technologies are very likely to create a solid return on investment. The following—partly overlapping—objectives were particularly prevalent in our case studies:

- More reliable scanning
- Better tracking
- Better tracing
- Integrated metadata management
- Reduced back-end communication
- More efficient label management
- Improved cooperation along the supply chain

We now discuss these areas in turn.

1.3.1 Scanning

A comparison between RFID and a conceivable or existing barcode solution is typical for many situations. RFID technology, while often more expensive than barcode solutions, offers a number of substantial advantages that make it worthwhile to consider a switch. First, RFID tags can be read without obtaining a line of sight. This may make it much easier, especially in a manufacturing environment, to implement an efficient scanning process. Some objects in the manufacturing plant may be shaped in a way that makes it hard for barcodes to be attached so that they are always readable. Optical barriers occur frequently on the shop floor. This always leads to requirements for manual interaction, while RFID reading can be automated completely. Second, RFID tags allow for bulk reading. Large numbers of objects can be scanned virtually at once, whereas barcode scanning works only on an object-per-object basis. Third, plant floors of manufacturing companies are often hostile environments with extreme conditions. Dirt, heat, the presence of metal, as well as limited space pose challenges that may be impossible to match by barcode- and network-based solutions. RFID tags have the advantage that they can be covered in protective casings and may thus be more reliable than barcodes. Moreover, writable tags do not depend on a network (see Sect. 1.2.6).

1.3.2 Tracking

As experiences from the retail industry have shown, the RFID-based tracking of parts, devices, and containers can lead to improved shipment and inventory management. With the availability of high-quality comprehensive data about the production process, labor costs can be reduced, and the overall process can be accelerated.

Cost depends on the hardware implementation. To ensure tracking on a shop floor, reader gates would be required at waypoints on the transportation routes. Additionally, RFID tags would be needed to label the model parts. Alternatively, a company could install mobile readers and position tags on the plant floor. Higher-class RFID tags can collect sensor data about physical conditions of the production environment. The autonomous energy supply and the ability of sensor tags to form ad hoc mesh networks allows for easy deployment in production environments.

Better tracking helps ensure accurate and real-time reporting about production status. Achieving compliance, narrowing recalls, addressing liability issues, and receiving better data about the production processes are among the possible motivations (see Sects. 4.4–4.6).

Typically, a production step is booked into the MES after the processing is completed. Information about the process status is thereby fed into the back-end system. This information may also be required in future consistency checks and used for later process analysis. During the case studies, it was common for manual booking to be sometimes forgotten or not conducted in a timely fashion (see Sects. 4.2 and 4.5). Possible consequences are inaccurate data tracks, wrong status information, and even interruptions in the production process. These problems can be overcome if RFID tags are applied on the materials or on the internal transportation units. In such setups, RFID readers could automatically detect whether materials are transported to the next process step, and consistency would thus be ensured.

Improved tracking also allows for more thorough process analysis, which may help reduce production errors and improve product quality. This could particularly be achieved by automating the production-related data management and thereby avoid errors in manual data maintenance. The frequency of errors and the related costs determine the potential savings for this use case. Furthermore, improvements in data accuracy may help in analyzing the processes better and identify potentials for improvements. Event tracking may also lead to a swifter response to problems and emergencies, thus minimizing downtimes. Realizing an emergency system with RFID would require rewritable tags with a few kilobytes of memory as well as one reader per workstation. These investments must be compared to the probability of downtimes of the backend system multiplied by the average cost of such an incident.

Some customers of the companies we surveyed require process reliability and process documentation (for instance, consistency checks to ensure that no process step is skipped). In an example from the case studies, a company scanned its products after every processing step in order to be compliant with customers' demands (see Sect. 4.1). RFID can help meet such demands considerably.

1.3.3 Tracing

A key driver for implementing RFID is expected savings from improving the traceability of products, especially in case of failure. If a production error is detected, all potentially affected parts must be checked manually. For faulty and already shipped products, checks must even take place at the customers' plants. This results in additional costs for sending workers to the customers. Costs for recalls are even higher for parts that where already used in a customer's production process; in this case, the manufacturer may have to pay penalties. Therefore, improving traceability and thereby narrowing recalls can account for significant savings.

Our case studies confirmed that recalls can be a major cost factor for production plants. Consequently, narrowing recalls to the products that were actually affected is desired, so detailed data tracks and process information are necessary. Exact information about which object was manufactured out of which components and materials is required to identify all products that include potentially flawed parts.

In addition, fine-grained and reliable data records can be important in legal disputes (see Sect. 4.1). A company may be sued if malfunctioning products cause damage. In this case, data records are important to prove that the production was conducted correctly. Here, sensor data can help detect the cause of failure and further narrow the scope of potentially affected products. Evaluating this data can provide additional insights into performance measures such as cycle times. It can even help identify the cause of quality changes.

1.3.4 Metadata Management

Accompanying documents are currently a frequently used way to maintain data in the production line. These paper documents are transported along with their corresponding materials and are used to record (meta-) data about the production process. Additionally, these documents can hold information about how to conduct subsequent processing steps. The accompanying documents are usually only loosely coupled with the objects they belong to. That is, documents move along with the corresponding objects but are physically separated from them while data are written on the paper. This may cause a mix-up of documents and lead to incorrect data maintenance.

RFID tags with writable memory can be used to store metadata about relevant events and processes right on the corresponding object. This way, information cannot get lost or accidentally mixed up. This solution may also be leveraged to automate some of the manual data maintenance. For instance, records of the conducted processing steps can automatically be written from machines to the tags. Automatic data maintenance would account for time savings and a reduction in mistakes.

Users should be aware that information could also be stored automatically in the back-end system. An identifier on the corresponding object would be needed, such as a barcode label or an RFID tag. Exchanging such data with the back-end system requires an appropriate network infrastructure and software system. The investment in such an infrastructure must be traded against the investment in RFID tags and readers, and the resulting network load must be considered carefully.

1.3.5 Back-End Communication

In contrast to barcodes, RFID tags can store up to several megabytes of data. More complex tags even have processing capabilities and programmable logic. These technical properties enable a novel distribution of processing logic in the IT infrastructure. Specifically, it is possible to shift the execution of business logic from monolithic back-end systems to the edge computers or even the tags themselves, thereby decoupling the processing logic from the back-end system. However, tags that store more than just identifiers are relatively expensive and may therefore be primarily applied in cases where they can be reused.

Some of the investigated companies expressed the demand to reduce interaction with the back-end system. In one case, the network infrastructure and the back-end database were perceived as unreliable (see Sect. 4.6). Consequently, IT support for production should work without the back-end system (at least in an emergency in case the back-end system fails). In another case, the company's network and back-end computers were hardly able to serve the demanded response time (see Sect. 4.1), and the IT staff has predicted significant bottlenecks when data volumes increase in the future. In both cases, RFID tags with writable memory could help decouple the processing of business logic from the back-end system. Currently, interaction with the core back-end system is needed to retrieve task-related data at each processing step. These data could also be stored at the object of interest if RFID were used. In this setup, edge servers that read the RFID data may execute business logic without consulting the back-end system. As a result, data retrieval would speed up, and the back end would no longer be a single point of failure.

1.3.6 Label Management

The case studies confirmed that manufacturers face challenges in handling labels at the outbound shipment; different customers demand different barcode solutions for labeling transportation units and packages. These differences can be in the demanded label format, coding scheme, or information on the label. Another challenge is that customers may claim financial compensation from the manufacturer if barcode labels are unreadable. In a specific example from the case studies, a customer's production line stops if a barcode happens to be unreadable (see Sect. 4.1). The resulting costs must be reimbursed by the manufacturer that printed the barcode.

RFID tags with writable memory can hold data in arbitrary coding schemes. RFID may thus be leveraged to abstract from the physical level. Different data amounts and coding schemes would no longer require different label formats. That is, one kind of RFID tag could be used for all customers. Label handling could be further improved by increasing the labels' readability. This could at least be done in environments where dirt or mechanical influence can affect the barcode. However, for this use case, all customers would need to accept the RFID tag protocol being used.

If RFID readers are in place, they can be used to write customer-specific information on tags in the outbound. In this case, expensive, specialized printers for labels would no longer be necessary. Label printing is currently done centrally to reduce the number of costly printers. If RFID readers could be used instead, central printing could be avoided, and the process could be organized more efficiently. This would result in savings on hardware costs and increased productivity.

1.3.7 Inter-Enterprise Collaboration

Partly as a result of the potential improvements described above, the introduction of RFID in manufacturing may improve the cooperation of enterprises along the supply chain. The key for successful collaboration is a controlled sharing of information that works to everybody's advantage ("optimal transparency"). The information can be transferred on the RFID tags that traverse the supply chain, or it can be held in one or several back-end repositories. More important than the technical realization, however, is the question of how to ensure that the efficiency gains are fairly distributed among the various participants. If this cannot be achieved, some enterprises may oppose the introduction of RFID simply because the cost outweighs the benefits. In this case, partner companies may want to establish compensation payments in order to avoid a "prisoner's dilemma" in which the efficiency of the supply chain remains suboptimal because of the opposition of only a few participants. We will discuss this in more detail in the following section.

1.4 Cost–Benefit Considerations and Adoption Decision

The RFID application areas discussed in the previous section can be categorized along two dimensions, as visualized in Fig. 1.9. We distinguish between operational use and strategic use on the one hand and between intra-enterprise and inter-enterprise applications on the other hand.

The term *operational use* refers to improvements that impact processes and productivity directly. That is, the RFID technology is applied to make processes faster, more secure, and so on. Furthermore, improved activity planning and better resource allocation due to RFID falls into this category. The term *strategic use* refers to use cases in which RFID is introduced for long-term strategic purposes. For instance, RFID may allow a company to provide additional services or quality guarantees to customers, thus changing the market positioning of the enterprise as a whole. Strategic decisions have a long-term impact, whereas operational decisions focus on immediate results.

Use cases for RFID in which the technology as well as the captured data are used only within one organization fall into the category of intra-enterprise applications. However, RFID holds potential for applications that span several steps in the supply chain. For instance, enterprises may exchange data that are obtained using RFID. Also, RFID labels may remain on products and could be reused by several companies downstream in the supply chain. Such use cases are categorized as inter-enterprise applications.

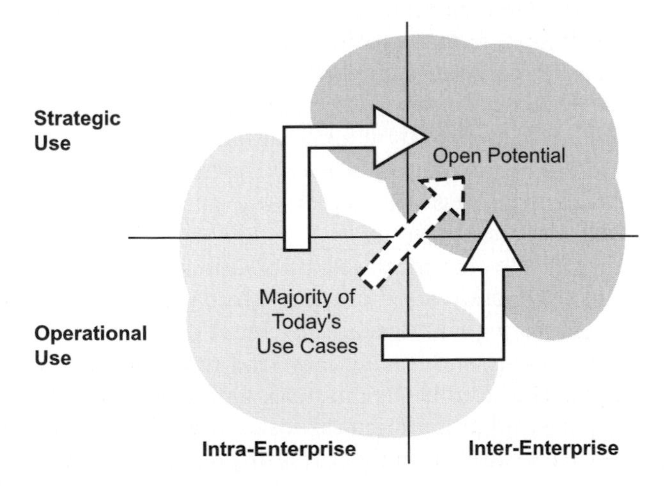

Fig. 1.9 Taxonomy of motives for RFID adoption

During our case studies we found that most RFID use cases today fall into the category of operational, intra-enterprise use. That is, the technology helps improve processes and productivity on the plant floor. Use cases in which RFID is used as a strategic enabler are found much less frequently. With regard to planning, RFID-enabled decisions are also mainly of local scope. In most cases, only locally obtained RFID data are used, and decisions are made for local processes.

The current focus on operational, intra-enterprise applications neglects the potential for RFID technology to be used in many steps of the supply chain. Operational use across enterprises can be enabled by the reuse of RFID tags. One option is to permanently leave RFID tags on the products as they move through the supply chain. Another option is to reuse the tags in a closed loop but extend the loop across several enterprises. This not only supports seamless integration of processes but also enables cost-sharing models for hardware expenses. Beyond this, RFID use across enterprises may strengthen the strategic position of the supply chain as a whole. For instance, cooperating partners can use the technology to provide fine-grained product traceability and quality assurances across the whole supply chain. Moving toward this opportunity can become a distinguishing factor and competitive advantage for innovative manufacturers in the near future.

We anticipate for the near future that manufacturing companies will introduce RFID with the expectation of realizing short-term benefits in the lower left quadrant of Fig. 1.9—that is, through operational, intra-enterprise measures. From there, we expect many companies to move gradually to the upper right quadrant, often by traversing the lower right quadrant or the upper left quadrant first.

RFID holds considerable potential for improving relevant business indicators. The cost of an RFID implementation includes some costs for hardware and software that are relatively easy to measure. Depending on the required functionalities, tag prices may range from about 5 cents to several euros. Reader prices range from a few hundred to several thousand euros. Tags may be reused (closed-loop applications) or remain attached to objects. However, there are costs that are notoriously difficult to measure, including system integration costs and costs of adapting business processes. In particular, as our case studies have shown, a successful RFID introduction greatly depends on the IT infrastructure already in place. This affects important architectural and functional aspects such as integration with other systems (S95 level 3 functionality), paradigms for data and event stream processing, distribution of logic and data, and support for heterogeneous data sources.

Concerning benefits, RFID may lead to increased automation, especially in data capture, and therefore to a reduction of labor costs. More important, the introduction of RFID may lead to a qualitative improvement of the relevant business processes. Improved tracking and tracing may lead to a more stable manufacturing process, with interruptions in the production process becoming less frequent. Tracing faulty parts and processes in the wake of a complaint or an accident is becoming much easier, and given the increasing demands regarding product liability, this is likely to create major competitive advantages for early adopters. In container management, RFID can reduce the overall cost of purchasing and renting containers. Tracing containers would allow reduction in the safety stock of containers, and fewer containers would need to be rented. Furthermore, loss of containers could be reduced, or external partners could be held responsible for losses at their sites. Using RFID for the uniform labeling of shipments (label management) could also lead to considerable savings in labor and hardware. Concerning process safety, production waste could be reduced by avoiding false machine settings. The resulting savings of this use case depends on the value of the wasted material and the cost of processing it.

As discussed above, a major problem when considering RFID introduction in a supply chain is that costs and benefits are not always correlated. Some participating companies may incur considerable costs that outweigh the local benefits, and vice versa. This can lead to a classic prisoner's dilemma: It could well be possible that an existing supply chain could gain considerably from introducing RFID technology. These gains, however, are never realized because some participants would need to incur costs that are not justifiable in comparison to their local benefits. As a result, these participants decide—for completely rational reasons—not to adopt the new technology. One way to break this deadlock is to negotiate compensation payments between different participants in the supply chain, with the objective of distributing the global benefits fairly among the participants. These compensation payments do not have to be monetary; in the retail domain, certain types of data, such as sales data about one's own products or the products of one's competitors, are also common currency.

1.5 Summary

The purpose of this chapter was to lay the foundations for the upcoming discussion of RFID in manufacturing. A key observation concerns the fact that RFID develops its full potential only if it is tightly integrated with any existing IT infrastructures. A tight integration with existing ERP and MES systems makes it particularly more likely that RFID will lead to concrete and local productivity improvements in the short and medium terms. Such concrete improvements—especially if they are purely intra-enterprise, i.e., independent of any coordination with supply chain partners—facilitate the adoption decision considerably. Cost–benefit calculations are less complex, and there is less uncertainty about the medium-term profitability of an investment in RFID technology.

We suspect that many enterprises will first focus on operational, intra-enterprise applications of RFID. Several examples of such applications were given in this chapter. We also pointed out the importance of working standards for both hardware and software. Even intra-enterprise applications rely on well-accepted standards to integrate hardware and software from different vendors.

Once an enterprise has introduced RFID on the shop floor and obtained tangible productivity improvements as a result, we expect more and more companies to look at the strategic potential of the technology. Inter-enterprise applications will become increasingly popular, even though they require close cooperation of the supply chain partners. In many cases, detailed negotiations will be necessary in order to distribute the costs and benefits fairly.

In the following chapters we continue with a presentation of the salient features of MES and ERP systems and show how RFID technology can be integrated into existing ERP- and MES-based IT infrastructures.

Chapter 2
The Role of Enterprise Software

Coauthored by Christof Bornhövd, Gregor Hackenbroich, Stephan Haller, Tao Lin, Jochen Rode, and Joachim Schaper[1]

2.1 Trends in Manufacturing and ERP

In Chap. 1 we gave an overview of how an IT systems landscape should look according to the ideal, simplified three-layer architecture. But this strictly layered approach is no longer feasible in modern IT environments. Modern business trends ask for more flexible, service-oriented architectures that allow, for instance, an ERP system to directly interact with an RFID reader on the shop floor. Using SAP software as an example, we will illustrate this type of modern architecture and the core technologies for real-world awareness, including RFID enablement.

To stay competitive in a global market, manufacturers must be able to react quickly to changes in their business environments. A global survey (Franklin 2005) conducted by the Economist Intelligence Unit of 872 executives in the manufacturing industry showed that they see adaptability to a changing environment as one of the key challenges for their businesses during the next couple of years.

Changes in the business environment are accelerated through new trends up and down the supply chain. Downstream, customers ask for more individualized products, fostering trends toward mass customization and smaller lot sizes. Customers also no longer accept extensive order times and want to be able to change orders as late as possible (late-order freeze).

Upstream, manufacturers depend heavily on their supply network and have to know as soon as possible about unforeseen delays in the delivery of their supply parts. This dependence has steadily increased during the last couple of years because the pressures for cost reduction and lean manufacturing prohibit large warehouses that could buffer delays. Instead, just-in-time and just-in-sequence delivery of supply parts is very common.

---

[1] This chapter is partly based on Bornhövd et al. (2004) and Hackenbroich et al. (2005).

O. Günther, W. Kletti, U. Kubach, *RFID in Manufacturing*
DOI: 10.1007/978-3-540-76454-0, © Springer 2008

These trends within the product value chain ask for a horizontal integration between business partners that on the one hand is more flexible but that, on the other hand, provides better and faster information flow between the partners. RFID technology has been proven to truly help meet these partially competing goals, especially in supply chain management, such as by providing near real-time information about the shipment status of supply parts. The increased capability of manufactures to adapt to potentially unforeseen changes in their upstream and downstream business processes is often referred to as adaptive manufacturing, which might go as far as the ad hoc outsourcing of parts of their production.

To keep pace with the highly dynamic environment of the manufacturers, their IT systems and application software also had to become much more flexible and cannot be as monolithic as they were 10 years ago. Service-oriented architectures (see Sect. 2.2) have been a big step forward on the way to providing manufacturers with application software that is flexible enough to handle their rapidly changing business processes and their dynamic business environments.

In addition to the horizontal integration, vertical integration across the ERP, MES, and automation layers plays an increasingly important role for manufacturers. Shorter production planning cycles and the need for even more efficient production plans call for more detailed and timely information about the status of resources on the shop floor. Short-term customer requests demand for available-to-promise (ATP) checks within minutes. Also in the context of this vertical integration, RFID is an important technology that can provide additional information about status and in particular about the location of shop floor resources.

Like the need for improved horizontal integration, the trends toward a stronger vertical integration are reflected in changes in manufacturing application software. New technologies have been developed that directly connect ERP systems to shop floor equipment, RFID readers, and other hardware devices (see Sect. 2.4). Functionalities such as quality management and detailed scheduling are being reassigned among the various software components, and the delimitations between ERP, MES, and specialized control software are gradually becoming less sharp.

In the following sections we give an overview of how modern software architectures meet the needs for more flexibility and improved information flow within the company and across its borders, using SAP software as an example.

2.2 Enterprise Service-Oriented Architectures

Instead of delivering monolithic enterprise applications consisting of only a few modules, software companies nowadays are in the process of switching to service-oriented software architectures. In such architectures, applications are built by combining various services. Each service provides specific functionalities through standard Web interfaces. This even makes it possible to combine services from different vendors.

With its enterprise service-oriented architecture (enterprise SOA), SAP already delivers such a software architecture. The foundation of this architecture is a new business process platform. It consists of the SAP NetWeaver technology platform and an enterprise services repository that includes the individual enterprise services. Current and all future SAP applications will be built on top of this platform. The platform also provides the possibility for independent software vendors to offer their own services on top of the business process platform.

With the SAP Composite Application Framework (SAP CAF) tool, SAP's own services and those of independent software vendors can easily be combined to new applications. Some of them, referred to as SAP xApps composite applications, are sold and supported as products of their own. Examples of SAP xApps include SAP xApp Product Definition (SAP xPD) and SAP xApp Resource and Portfolio Management (SAP xRPM). Currently, more than 150 SAP xApps are available.

Among others, the SAP NetWeaver platform includes the following key components:

- SAP NetWeaver Application Server: a powerful application server
- SAP NetWeaver Exchange Infrastructure (SAP NetWeaver XI): a robust integration engine for both application-to-application and business-to-business scenarios
- SAP NetWeaver Business Intelligence: a high-performance business intelligence engine
- SAP NetWeaver Portal: a flexible and customizable portal for information delivery to users both inside and outside the corporate firewall
- SAP NetWeaver Master Data Management: a system for harmonizing information that is distributed across a wide variety of applications

2.3 SAP Software for Real-World Awareness

SAP offers a number of purpose-built applications to link sensor and event information from the "real world" to SAP's business process platform and enterprise applications. These include SAP Auto-ID Infrastructure and the SAP xApp Manufacturing Integration and Intelligence (SAP xMII) composite application. Both SAP Auto-ID Infrastructure and SAP xMII are often deployed in a distributed environment, with servers located close to the devices in a warehouse, distribution center, or production facility.

SAP Auto-ID Infrastructure provides prebuilt functionality for common RFID integration scenarios in supply chain activities, open infrastructure for device controller integration based on ALE, and open enterprise and inter-enterprise connectivity based on the EPCglobal™ standards. With SAP Auto-ID Infrastructure, SAP has achieved an important milestone in realizing its vision of an adaptive supply chain network. (More details are provided in Sect. 2.4.)

SAP xMII provides a rich framework for integrating a wide range of automation, information, and sensor systems found in manufacturing plants, utilities, warehouses, and other locations where application-to-device integration is needed. It supports relevant device and application integration scenarios such as OPC and S95/B2MML.

Many RFID applications need to interact with other equipment such as conveyors, material handling systems, weigh scales, programmable controllers, and many other types of devices. SAP xMII provides this link to the data and events from these sources, as well as a means to write to these devices.

In addition, SAP provides tools for building mobile-specific applications to deal with the unique needs of mobile workers, occasionally disconnected scenarios, and mobile devices such as forklift-mounted terminals, personal digital assistants (PDAs), and other mobile computing hardware.

The combination of these key enablers for "real-world awareness" along with SAP's enterprise applications and powerful SAP NetWeaver platform creates a comprehensive set of software components for dealing with almost any RFID-based application requirement. These solutions enable the linking

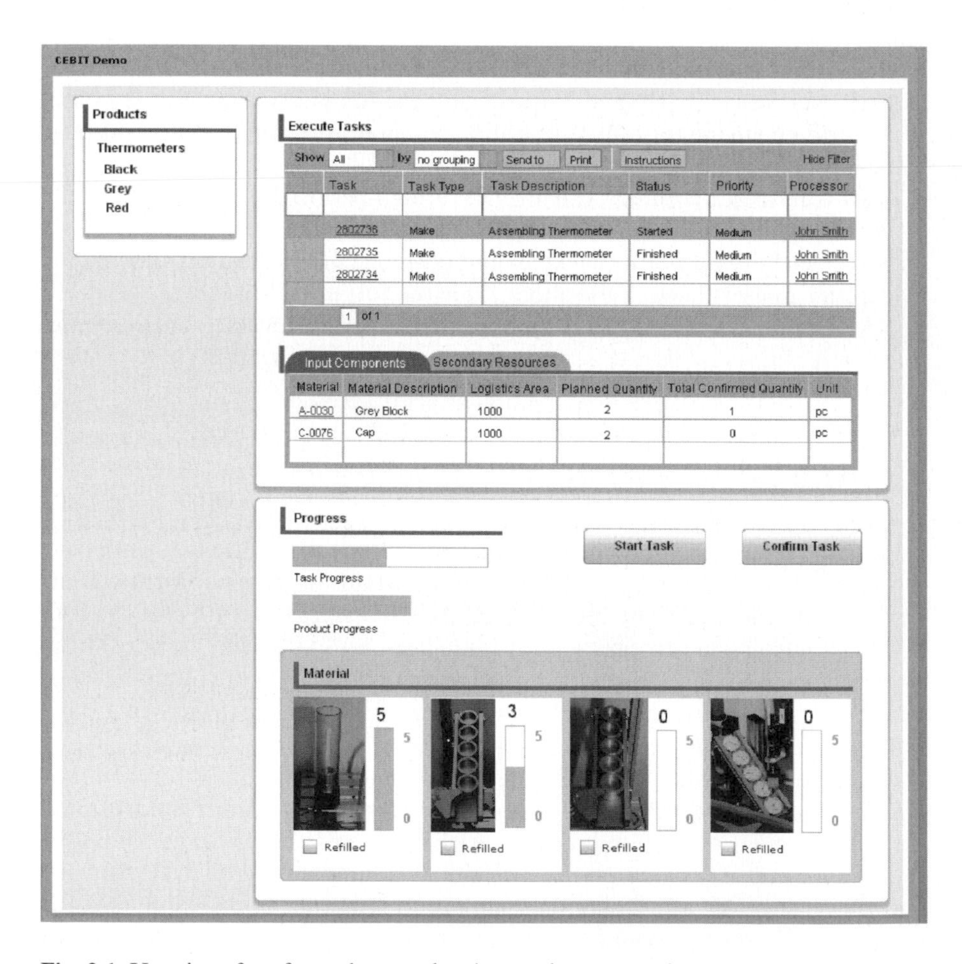

Fig. 2.1 User interface for a plant worker (research prototype)

of line-of-business applications with external events such as those provided by RFID, as well as provide mechanisms for integrating with other critical sensor data and with automation systems, thus "closing the loop" for automated business processes.

In many cases, the "biological interface," the worker, is the source or destination for information or transactions, and the SAP family of products addresses the user-interface requirements as well. Using the portal technology mentioned above, it is easily possible to integrate very heterogeneous sources of information. Figure 2.1 presents a prototype (developed by SAP Research, SAP's global research organization) of a user interface for plant workers. It shows information about the tasks the worker is currently working on; this information is extracted from the business process platform. In addition, the user interface provides information that comes directly from the shop floor equipment, such as task progress and inventory data. The information is read via SAP xMII and OPC from the shop floor equipment. The production tasks are triggered through RFID tags that are read through SAP Auto-ID Infrastructure.

2.4 SAP Auto-ID Infrastructure

2.4.1 System Requirements

SAP Auto-ID Infrastructure was architected with the following system requirements in mind:

Scalability: Companies like large retailers are assumed to require throughput rates of about 60 billion items per annum within their supply chain. Assuming 100 distribution centers, each with an average of five checking points per item, the system needs to guarantee an average throughput of at least 100 messages per second per distribution center. The size of an observation message can be assumed to be around 200 bytes, and the processing of an incoming observation message usually requires multiple database updates and the execution of business procedures at the back-end system.

Open system architecture: In addition to being hardware-agnostic, the architecture should be based on existing communication protocols such as TCP/IP and HTTP, as well as syntax and semantics standards such as XML, PML (Floerkemeier et al. 2003), and EPC (EPCglobal 2005). This allows the use of sensors from a wide array of hardware providers and will support the deployment of auto-ID solutions across institutional and even national boundaries.

Efficient event filtering: The infrastructure needs to provide efficient means to filter out false or redundant readings from RFID or sensor devices. Also, it needs to provide flexible and configurable filtering of events to pass on only relevant information to the appropriate back-end processes.

Event aggregation: The infrastructure needs to support the composition of multiple related events to more complex events for further processing. For example, in supply chain applications the system must allow the composition of individual object identification events for multiple individual cases and the corresponding pallet to only one complete-pallet-detected event.

Flexibility: The infrastructure needs to be adaptable to different business scenarios, for instance in manufacturing or supply chain management. Furthermore, the infrastructure needs to provide a flexible means at the business logic layer to respond to abnormal situations, such as the missing of expected goods or company-internal rerouting of goods. To avoid redundant implementations of the same business rules in different enterprise applications, the infrastructure needs to offer a way to deploy and execute them within the auto-ID infrastructure.

Distribution of system functionality: A real deployment of an auto-ID solution can be distributed across sites, across companies, or even across countries. This naturally requires a distributed system architecture. As a first step, we require that the auto-ID infrastructure support the distribution of message preprocessing functionality (for example, filtering and aggregation) and, to some degree, business logic across multiple nodes in order to better map to existing company and cross-company structures.

System administration and test support: The infrastructure must provide support for testing individual custom components used in the filtering and aggregation of events, as well as the end-to-end processing of RFID and sensor data. Good administration and testing support is a prerequisite for the deployment of a distributed auto-ID solution in large-scale applications.

2.4.2 System Overview

The architecture of SAP Auto-ID Infrastructure is shown in Fig. 2.2. Conceptually, it can be divided into the following four system layers:

At the *device layer*, different types of sensor devices can be supported via a hardware-independent abstraction layer. It consists of the basic operations for reading and writing data and a publish/subscribe interface to report observation events. By implementing this application programming interface (API), different kinds of smart item devices can be deployed within the auto-ID infrastructure. Besides RFID readers, these devices can include environmental sensors or PLC devices.

The *device operation layer* coordinates multiple devices. It also provides functionality to filter, condense, aggregate, and adjust received sensor data before passing it on to the next layer. This layer is formed by one or more device controllers (DCs).

The *business process bridging layer* associates incoming observation messages with existing business processes. At this layer, status and history information of tracked objects is maintained. This information includes object location, aggregation information, and information about the environment of a tagged object. A so-called auto-ID node realizes this functionality.

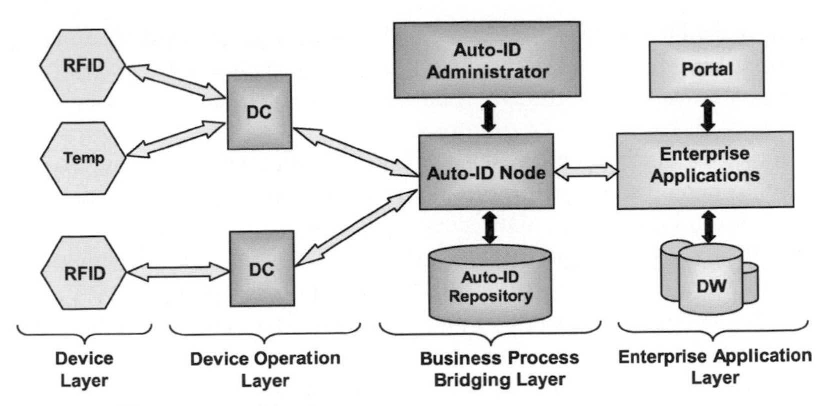

Fig. 2.2 Auto-ID system architecture

Finally, the *enterprise application layer* supports business processes of enterprise applications (such as those for supply chain management or asset management) running on SAP or non-SAP back-end systems.

SAP Auto-ID Infrastructure provides an infrastructure for realizing a complete auto-ID solution. Because auto-ID solutions can span organizations and even countries, standards for the interfaces between the components are essential. Therefore, SAP Auto-ID Infrastructure is compliant with the standards proposed by the EPCglobal™ consortium.

As part of the infrastructure, a test and workload generator tool is provided that can simulate messages coming from one or more DCs or back-end systems to an auto-ID node. Also, a scriptable simulator is available that can simulate multiple RFID readers. These tools allow the testing of an auto-ID deployment without the installation of physical devices.

The following two subsections explain the two main building blocks of the SAP Auto-ID Infrastructure: the device controller (DC) and the auto-ID node.

2.4.2.1 Device Controller

The DC is responsible for coordinating multiple smart item devices and reporting incoming observation messages to one or more auto-ID nodes. A DC supports two operation modes. In the synchronous mode, the DC receives messages from an auto-ID node for direct device operations, such as to read or write a specific data field from/to a tag currently in the range of an RFID reader, or to read the value from a temperature sensor at a given point in time.

In the asynchronous listening mode, the DC waits for incoming messages from the sensor devices. Upon receiving such a message, additional data can be read, and event messages can be filtered or aggregated according to the configuration of the DC. Note that when a DC is configured for asynchronous operations, it is still capable of synchronously receiving and executing commands.

Message processing in the DC is based on so-called data processors. We distinguish six different types of data processors:

1. Filters filter out certain messages according to specified criteria. For example, they can be used to filter out all event messages coming from case tags or to clean out false reads ("data smoothing").
2. Enrichers read additional data from a tag's memory or other devices and add this data to the observation message.
3. Aggregators can be used to compose multiple incoming events into one higher-level event (for example, mapping data from a temperature sensor to a temperature-increased event), or they can be used for batching purposes.
4. Writers are used to write to or change data on a tag or to control an actuator.
5. Buffers buffer event messages for later processing and/or keep an inventory of tags currently in the reading scope of an RFID reader.
6. Senders transform the internal data structure of the messages to some output format and send them to registered recipients.

The core functions of the DC, in particular the message processing described above, are independent of the hardware used. For reading and writing the data on the tags, we use logical field names to abstract from concrete tag implementations. A field map provides the mapping between memory addresses on the tag and logical data fields.

Because all data processors implement the same publish/subscribe interface, they can be arranged into processing chains. Powerful message processing and filtering operations can be achieved by chaining together the appropriate, possibly customized, set of simple data processors. This results in a very flexible framework that allows for the distribution of message processing functionality close to the actual sensor devices to reduce message traffic and improve system scalability.

Figure 2.3 shows an example of a typical processor chain used for dock doors in a supply chain scenario. For full coverage, dock doors commonly use more than one reader. This holds especially true in Europe, where there are much stricter radio frequency regulations than in the United States. RFID readers sometimes generate false event messages. For example, because of physical reasons a tag may not be seen during a particular read cycle. To filter out these false "tag-disappeared" messages, a LowPassFilter is applied. Also, every tag that passes the radio field will issue two event messages: a tag-appeared and a tag-disappeared message. Because in the dock door scenario one is interested only in the fact that an item has passed the door, one can safely filter out tag-disappeared messages by using an EventTypeFilter. The EPCEnricher in the example is needed only if non-EPC tags (which are still common today) are used. These tags have a unique ID that is set by the manufacturer, and the EPC is actually stored in the tag's user memory. In this case, the EPCEnricher reads the EPC and adds it to the event message. At a dock door, one wants to collect

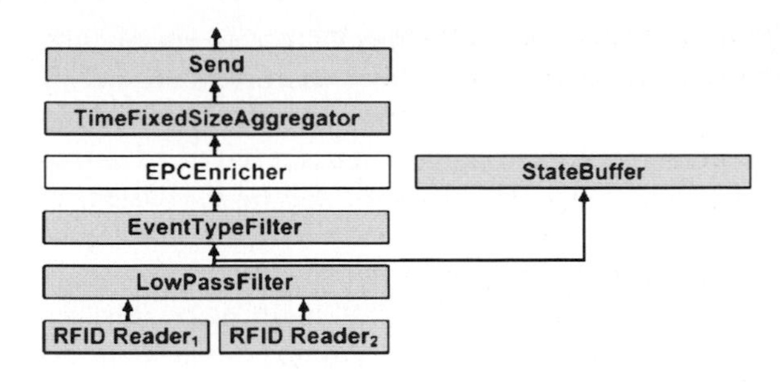

Fig. 2.3 Typical data processor chain

all tags that are seen during a certain time window and report them in a single message to the back-end system. The TimeFixedSizeAggregator and the Send processor in the example do this. In addition, a StateBuffer keeps track of all tags currently in the reader's scope for auditing and reporting purposes.

2.4.2.2 Auto-ID Node

An auto-ID infrastructure can contain multiple auto-ID nodes. An auto-ID node is responsible for integrating incoming observation messages from the DCs with the business processes running at the back-end systems.

For an auto-ID node, one must distinguish between the interactions with DCs (reader events from and control commands to DCs) and interactions with back-end enterprise systems (such as receiving master data from a back-end system and returning a confirmation). These interactions with the auto-ID node are treated as either incoming or outgoing messages.

Incoming observation messages are routed to a rule engine, which, based on the message type, evaluates a specified list of conditions. The result of the evaluation step is a set of qualifying rules for which one or more actions are executed in a specified order. Such an action can, for example, update the system status of an object in the local repository, communicate with the back-end system, or generate and write EPC data to a tag.

Actions of a rule can pass on parameters and can trigger other rules at the auto-ID node. Based on the message type, messages can be assigned different processing priorities and can be specified as being persistent in the auto-ID node.

An auto-ID node provides a local repository that contains information about the current status and history of the objects being processed. This information includes data about the operations that have been applied to an object (e.g., move, pack, or unpack), its movement and current location, and its structure (e.g., packing information). Also, the repository replicates master data from the back-end system about products and business partners or the physical location and type of the RFID readers. The auto-ID repository provides the basis for the execution of business logic in the auto-ID node.

The use of customizable rules provides a flexible mechanism to specify and execute business logic at the auto-ID node. This allows the preprocessing of incoming observation data and the handling of abnormal situations within the auto-ID infrastructure, such as discrepancies between a received advanced shipping notification and a detected pallet. This, in turn, allows the system to offload processing from the back-end systems.

SAP Auto-ID Infrastructure was initially developed to cover supply chain management scenarios, for which a standard set of rules is in place. The deployment of an auto-ID infrastructure in a different context, such as in asset management or manufacturing, simply requires adopting or extending the existing rules. An example for such an extension in the manufacturing context is the successful deployment of SAP Auto-ID Infrastructure in a kanban scenario with a large German car manufacturer.

2.5 Summary

In this chapter we first described the upcoming trends in manufacturing and ERP. The strict application of the classic three-layer architecture (illustrated in Chap. 1) is no longer feasible in modern IT environments. There is a clear demand for more flexible, service-oriented architectures that allow, for example, an ERP system to directly interact with an RFID reader on the shop floor. Therefore, in this chapter we described how a modern architecture and core technologies for real-world awareness can be designed to meet this new demand. We used SAP software as an example for our illustration of a modern architecture.

We began our description with a short introduction about enterprise service-oriented architectures. This was followed by a brief description of SAP software as used for real-world awareness. A number of purpose-built applications exist to link sensor and event information from the real world to SAP's business process platform and enterprise applications. These include SAP Auto-ID Infrastructure, which is also the most prominent one; SAP Auto-ID Infrastructure is SAP's solution for integrating RFID data. In this chapter we discussed the requirements that impacted the development of SAP Auto-ID Infrastructure and described core components of this infrastructure.

Chapter 3
The Role of Manufacturing Execution Systems

From the user's point of view, the RFID tag is a data carrier to identify material resources, orders, people, or other objects whose "movements" are to be recorded in production. An RFID tag or the use of an RFID tag on its own does not constitute a separate application on its own. RFID has this fact in common with other identification systems such as reed relay, barcode, and the automatic recognition of shapes and colors. Because of its particular characteristics, however, the RFID data carrier can easily be "filled" with information and data, and the automatic transport of such data and information is simplified considerably.

Manufacturing execution system (MES) has become a common term in recent years to describe IT solutions for production and related functions. When considering RFID, it is advisable to first check the solutions that are already available in an MES and verify how these solutions can be enhanced by RFID technology. For this reason, this chapter discusses the efficiency and options of an MES.

3.1 Why an MES?

Tomorrow's plants will not primarily be simple series producers anymore. In addition to manufacturing large numbers of pieces with little retooling and product changeovers, modern production companies also regard themselves as service providers offering versatile products in a manageable quantity to each individual customer. Flexibility, lead time, on-time delivery performance, and product diversity are the keywords describing this new approach to production. In the future, customers will take for granted that they will receive high-quality products on time and at a favorable price. Along with product features, these three factors will become important components when competing on the global marketplace.

O. Günther, W. Kletti, U. Kubach, *RFID in Manufacturing*
DOI: 10.1007/978-3-540-76454-0, © Springer 2008

Because of these requirements, a typical production process will no longer run for extended periods of time; it will be interrupted to retool machines and change products. In so doing, product life cycles will be shortened. The increasing pace of innovation will produce new product versions. The terms "turbulence" and "volatility" describe these processes. Initially, these two factors increase costs and render it more difficult to maximize production efficiency. (For more details, refer to Kletti 2007b.) Moreover, they support inadequate information management—a problem that is aggravated by unsuitable or outdated business processes. Consequently, customers face poor delivery performance and unsatisfactory product quality. Lead times increase within production and cause excessive stock levels. The result is an unnecessary capital tie-up. The list of negative effects resulting from turbulence and volatility could easily be continued.

Modern production companies can cope with these effects by means of transparency, reactivity, and economic efficiency. This article describes how an MES can help achieve these three important process characteristics within manufacturing and how RFID affects this situation.

The claims for more flexibility and transparency require identification of all objects and processes within the production cycle.

The objective of increasing economic efficiency of production is not new. It is a basic element of industrial production. The same applies to the need for more transparency and reactivity. In discussions of how to realize these goals in the new factory of the future, the production types that are affected by these aspects should be taken into account first.

There are three different types of production:

- *Discrete or job shop manufacturing:* In this situation, production orders consist of a series of operations. Discrete producers intend to have short, optimal transitions between the single production stages. They usually work with a complex set of operating resources, and they try to keep an optimal flow of orders through the process. The availability of intermediate products and the organization of these intermediate products in intermediate storage facilities is an important element of the optimization strategy. In job shop manufacturing, other important issues are resource availability and flexibility with respect to the production flow.
- *Process production, line, or mass production:* In process production, aggregates and machines are integrated into lines generally producing large quantities of a product. It is hardly possible to realize flexible changes in the production of orders. Furthermore, it is very important that systems run constantly without interruption. Because of the complexity of equipment, it is to some extent impossible to reschedule orders or production operations. As a result, detailed production scheduling has its own rules.
- *Make-to-order production or systems production:* In make-to-order production, extensive bills of materials are used, and work is often carried out in production islands or specially designed workshops. The islands can be characterized by a certain independence so that transitions that are not time-critical to some extent arise between these islands. Depending on the

production type, the scope of an MES system will vary. The more complex the vertical integration and the shorter the cycles for production scheduling, the further the MES functions reach into the field of activity of an ERP system.

3.2 Integrating MES into the Enterprise Information Infrastructure

As discussed in Chapter 1, manufacturing companies have similar hierarchical structures, often represented as a pyramid. In the technical literature, one usually distinguishes three levels (see Fig. 3.1):

1. Company management
2. Production management
3. Production level (automation layer)

It is reasonable to divide the levels of production management and production because single processes within a production company are focused in this context (see Fig. 3.2).

This structuring simultaneously goes along with an assignment of production processes. The processes must have IT support to be able to proceed as efficiently as possible in terms of economic business objectives. ERP or MES systems classically assume the functions of mapping and supporting the processes.

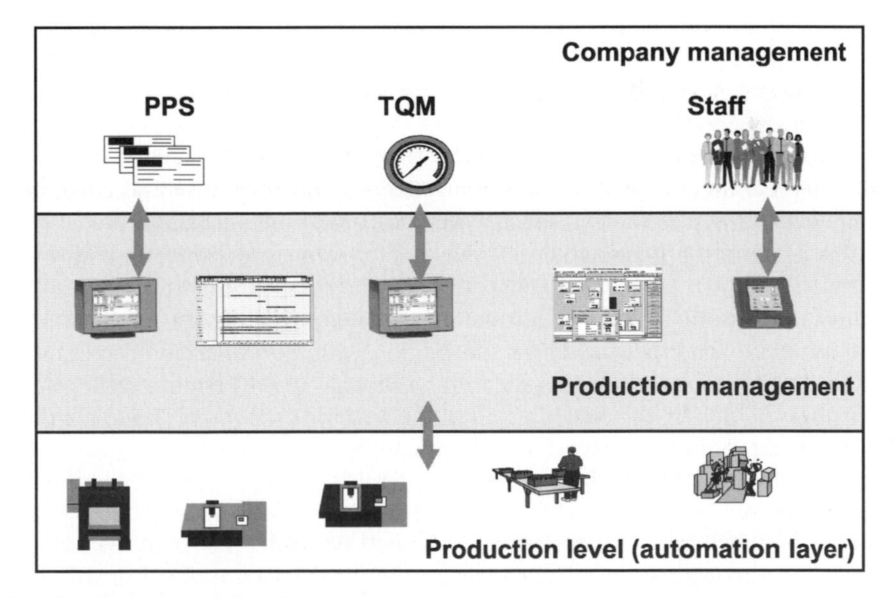

Fig. 3.1 Structure of a production company

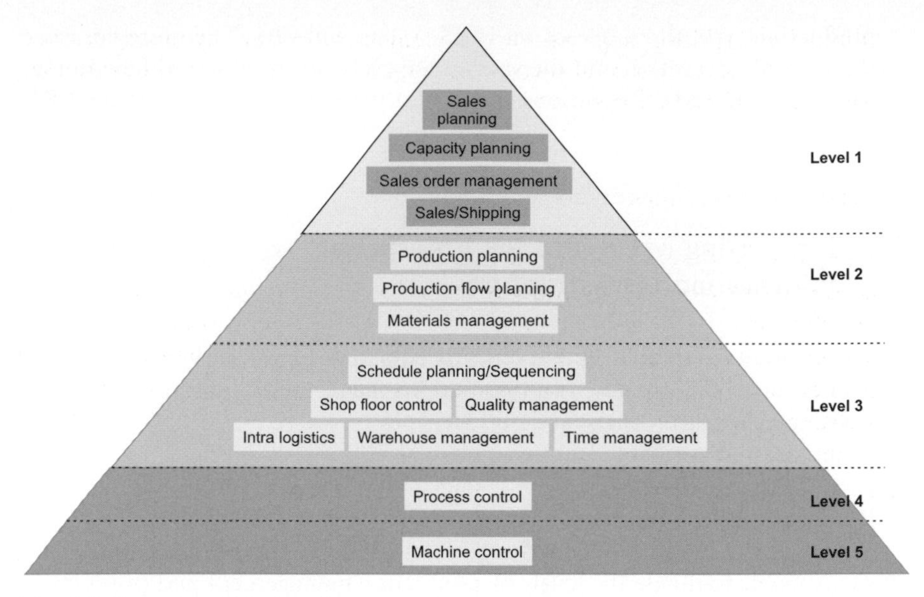

Fig. 3.2 Functional levels and processes in a company

Level 1: Company management

The typical ERP functions such as sales planning, capacity planning, sales order management, sales, and shipping are on the highest level.

Level 2: Planning and material requirements planning of the production

The functions for product planning, production flow planning, and materials management are on the second level.

Supply chain management also belongs to this layer, by means of which the complete logistic processing is controlled on the enterprise and corporate group level as well as the logistics between customers and suppliers. In contrast to this is transport management, which will be described below and which is commonly known as intralogistics, or the material control within the production of a company. The SAP Advanced Planning & Optimization component can also be found here.

The time horizon is basically medium or long term for planning processes.

Level 3: Production control

Level 3 includes all processes mainly designed to realize the plan, which was created on the superior level. The material flow and actual production are controlled in this layer.

Functions and tasks such as reactive planning, and thus the actual production control (also known as shop floor scheduling), can be assigned to this level, where a short-term time horizon is applied.

Level 4: Control of manufacturing process

Depending on the industry sector and/or production structure, functions that are largely assumed by process control systems in process industry/flow production can be found on level 4. In cases of discrete production, these functions for controlling the production processes are represented to a great extent on the production control level (shop floor scheduling). According to manufacturing organizations, quality management is also involved in this layer.

Level 5: Control of machines and equipment

All functions and processes of a production company that can normally be found on the level of machine or process control are located on layer 5. These functions are required to be able to control machines and equipment and also to exchange information and control parameters from and to the machine/aggregate.

3.3 Processes Relevant to MES in the Company

Processes that are important to the manufacturing environment and that provide the largest room for optimizations should be highlighted in this context.

These relevant processes can be assigned to an ERP and MES system with respect to IT (see Fig. 3.3). Moreover, it can be recognized that processes can run in the classic ERP environment as well as in the MES environment, thus in a kind of overlapping area of both systems. The following explanations are based on the SAP ERP application and the MES system HYDRA. Both systems provide production companies with a completely integrated solution regarding the system structure.

This assignment of processes is obvious for levels 1 and 5. The functions of level 1, business planning, are the typical tasks of an ERP system and are covered by SAP ERP. Production planning functions and material requirements functions are also mainly implemented in SAP ERP.

The functions of levels 4 and 5 are largely assigned to the MES system or at least must be able to be integrated into an MES system.

Production processes are controlled within the MES system (process control function). The classic data collection functions and the integration of

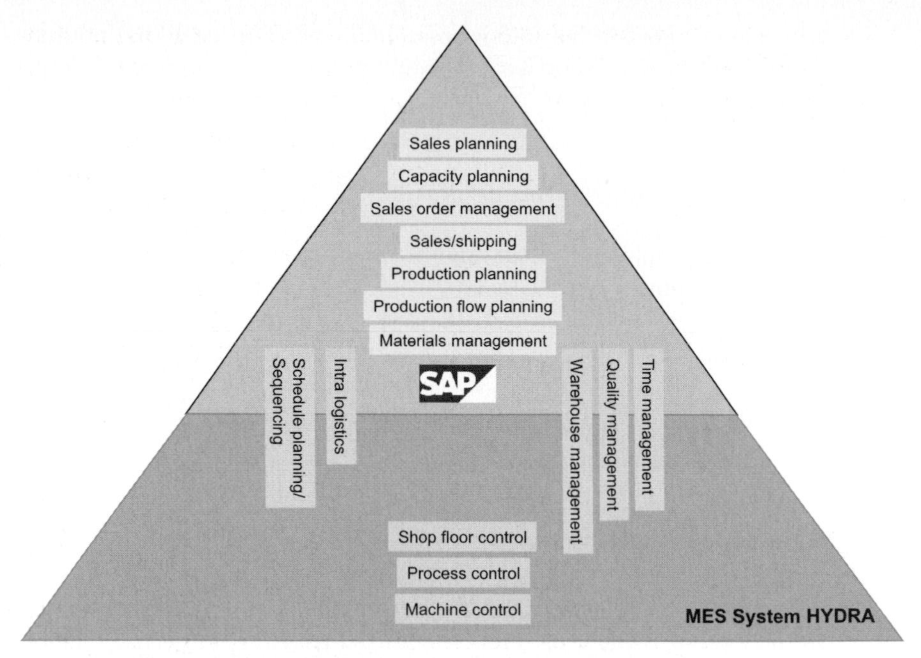

Fig. 3.3 Integration of a production company's processes into the IT environment

the relevant technologies to connect machines, aggregates, and process plants belong to this. The provision of all types of data for superior levels constitutes a fundamental prerequisite and is the basis to ensure requirements of vertical integration. (For further information, refer to Kletti 2007b.)

However, it has become commonly accepted that processes do not necessarily run completely in one system but have to be supported vertically beyond several system levels. So, along with the assignment of functions to a system, it has become the far more interesting part of system integration to "summarize" the functions from several systems to map one process.

As regards IT, this area of overlapping functions mainly affects the field of activity of level 3 in this model. According to the type of production (see Sect. 3.1) or industry sector, these processes can either be implemented completely in MES or SAP systems, or—and this seems to be the alternative of the future—they are resolved into subprocesses and are mapped in the SAP software as well as in the MES system.

As already mentioned, some functions/processes are either completely or only to some extent reproduced in the MES system. The connection of the MES system to SAP ERP has to be flexible enough that the following functions can be "activated/deactivated" in a company depending on the application and production organization. (For more information, refer to Kletti 2006, 2007a.)

3.3.1 Time Scheduling and Sequencing, Shop Floor Control (also refer to Sect. 3.7.6)

Although time scheduling and sequencing are done well in advance in terms of shop floor control (see also Sect. 3.7.6) in the production plan for series production, in some industries and manufacturing organizations these plans are either created just shortly before starting production or must be changed during production. To be able to differentiate between these planning types, the term "reactive planning" has been established. This term stands for the preparation of a practicable production plan that has been optimized with respect to production conditions at its creation. Reactive planning is required if there are technical restrictions, such as for different machines and production equipment due to the materials used.

The steel industry can be used as an example. Because of material properties planning, changes are determined directly before or during the production process, which might lead to overdeliveries or underdeliveries for the production order.

If other production stages requiring superior planning systems such as the SAP modules for production planning, materials management, or advanced planning and optimization are additionally affected, it is necessary to address them via the systems concerned. In this case, it is the task of the MES system to synchronously provide the SAP modules with information, to retrieve new planning specifications or even a new plan, or to trigger its creation.

3.3.2 Intralogistics (see also Sect. 3.7.3)

The notion of intralogistics (see also Sect. 3.7.3) may vary in meaning depending on the type of production and infrastructure of a single production company. Intralogistics does not generally imply the transport of materials from one production site to the next, to distribution centers, or even to the end customer. These processes are usually represented by the supply chain management functionality in the SAP system. But in addition to that, there are, for example, great distances that have to be bridged in certain production industries such as the steel industry. Within a plant, transport logistics between the individual production areas is an important task and completely independent from supply chain management. This fact also applies to manufacturing companies that have to face very difficult conditions regarding the infrastructure caused by growth over the preceding years (for example, parts of the company may be separated by public roads, or plant sections may be situated in the same city but with great distances between them).

In this context, transport logistics belongs partially or completely to the production process and has to be managed nearly as flexibly as the production process itself. Therefore, it is reasonable to map intralogistics as an individual logistics function within the MES system. The same applies to the typical work-in-process (WIP) materials that evolve during production. The intermediate materials resulting from this have only a short life cycle, so it is not necessary to keep them in a separate low-level code within materials management.

Besides the collection and tracing of intermediate materials in the production process, the documentation of the developing process of customer end products plays a decisive role on the transport management level. Batch tracing as well as individual tracing has become more and more important in the food and drug industry as well as in the automotive industry. In the areas of tracking and tracing, SAP provides increasingly more functions within its business solution. These functions, however, must be integrated into the production process by the MES system.

3.3.3 Quality Management (see also Sect. 3.7.4)

It has become a common fact for all industries that quality management (see also Sect. 3.7.4) is not an independent process in a production company but has to be integrated into the manufacturing process itself. This is known as in-production quality assurance. The employees working within manufacturing are supposed to check the quality of the product, adjust settings affecting the quality of the material or the finished product, or just collect quality data. The term "operator inspection" describes this fact.

To support this process, it is necessary that the data input functions at the machine or aggregate also contain monitoring functions for quality assurance. However, the control of this quality management and the archiving of the data lie within SAP quality management. The qualification regarding whether recorded data or processes are relevant for quality assurance or product documentation must be carried out directly within the production process.

The MES solution makes it possible for logistics management and quality management to run on the same level, which is required for in-production product documentation. As soon as data are generated, the relationship between product and quality data has to be established online by means of measured values because a connection of these data is impossible for some production processes at a later point in time.

3.3.4 Time and Attendance and Time Management Within Production (see also Sect. 3.7.5)

In a company, the functions from time management to the calculation of wages are part of the personnel management application called SAP ERP Human Capital Management.

In a production company, if incentive wages are to be paid to for the productive employees within manufacturing, this "rule" can be broken. In this case, it is important that data are preprocessed directly when they are generated and that the workflow for the data maintenance is ensured in time.

Example: If a foreman maintains the recorded data during the current shift instead of a time administrator doing it in the next shift or even on the next day, it is possible to compare the personnel and quantity records in time and to check them for plausibility. This leads to data quality and a secured production process as well as justified and, above all, efficient remuneration that complies with product quality and productivity.

For companies in which incentive remuneration does not play a role within production, it is reasonable to use time management in its classic meaning within the SAP environment (human resources module). Thus, the functions within the MES solution reduce to simply recording time events such as clocking in, clocking out, breaks, and such. But plausibility check requirements for all employees within manufacturing can also be solved via MES (see also Sect. 3.7.5). In terms of an overall solution, it is essential that the system (SAP software plus MES) support both alternatives.

Merely by describing the processes, it becomes apparent that a flexible identification system (such as RFID) might be very important, especially in intralogistics and quality management.

3.4 Tasks and Functions of an MES

Tasks that have to be fulfilled by an MES system derive from the coherences described so far. These requirements must be mapped, or "completed" within an MES system. Technologies such as RFID need to be an integral part of the MES functions.

Organizations such as the Manufacturing Enterprise Solutions Association (MESA), ISA, and Verein Deutscher Ingenieure (VDI; association of German engineers) have defined these functions by means of definitions, standards, and user-oriented guidelines. (For more details, refer to Kletti 2006, 2007.)

A list of eight tasks that the current VDI norm 5600 attributes to an MES solution constitutes a good summary:

Task 1: Detailed scheduling and detailed control

The detailed scheduling component of MES described in the VDI norm is considerably different from the ERP planning component that has medium-term and long-term effects. The essential aspect is control, which means that an MES detailed scheduling responds to a current situation within the production and thus updates an existing plan. An ideal MES mechanism does not always create new plans but carries out one single plan that can more or less be updated and optimized in real time.

Task 2: Management of operating resources

The term “operating resources” includes machines and equipment, tools, and other resources. The MES system registers the different times during which a resource is used and prepares respective statistics, by means of which the availability can be evaluated. The current status of a resource is transmitted to the other MES tasks online. Thus, the availability of resources can also be increased by considering preventive maintenance. By resource conditions that are provided online, the MES task of detailed scheduling and control is able to reschedule and optimize plans as necessary.

Task 3: Materials management

The materials management function administers WIP, arranges for the supply of workplaces with materials or components, and organizes the transport of the finished products. This MES function also keeps the status of materials and manages respective batches. Fully developed MES systems use this task as the basis for complete traceability solutions.

Task 4: Personnel management

Staff is the most important, or at least one of the most important, resources in a production company. The VDI guideline provides a separate MES task to use this resource efficiently. An essential issue is the effective scheduling and provision of personnel capacities, taking availabilities and qualifications into account. The personnel management task can also include resource scheduling and the keeping of time accounts. In this context, a particular task is to consider the legal general conditions based on which personnel capacities can be scheduled and used within a manufacturing company.

Task 5: Data collection

The real-time collection of all relevant data is of special importance within a production company. Useful operating conditions can be established only if data are recorded in real time. These operating conditions help perform efficient regenerative planning or rescheduling procedures. This is why data collection has a particular meaning in an MES system. It is differentiated into manual, semiautomatic, and fully automatic collection that also guarantees the support of machine and equipment interfaces. The recorded pieces of information are checked for plausibility, and the user is possibly made aware of errors. For this purpose, the data collection function has to know the master information required for the inspection. Preprocessing and compression of the data recorded allows for the formation of periods of time and quantities relevant for the operation from individual events. Other MES functions can use these periods and quantities as a basis.

RFID technology plays a decisive role on the data collection level.

Task 6: Activity analysis

Modern production companies remain competitive because the sum of many little improvements results in the great and necessary gains in performance that are required to be able to persist in international competition. Activity analysis, another MES function, continuously carries out a target/actual comparison, by means of which it can assess to what extent the production process lies within the required marginal productivity. This activity analysis is not performed in the form of rework or final costing as it is anchored within the way of thinking of ERP, but in terms of an online comparison between target conditions and measured actual values. Thus, it is possible to respond quickly and to correct or stop faulty processes.

An MES solution supporting this function should provide these data for a longer period and make a data basis available to influence processes in an organizational way. Especially, activity analysis provides quality management and other MES functions with its data. Moreover, activity analysis produces a series of known indicators such as the Overall Equipment Efficiency (OEE) index, key figures for resource utilization or processing times of orders.

Task 7: Quality management

In modern factories a very particular task is attributed to quality management. It is not only about manufacturing top-quality, functional products, but also about adhering to respective norms and regulations during the production of articles. Particular keywords in this context are traceability or, for food and pharmaceuticals production, compliance with the U.S. Food and Drug Administration (FDA). The MES quality management function provides support during quality planning, manages appropriate gauges, and carries out quality inspections. Consequently, quality management provides inspection documentation and certificates for all other MES functions concerned with information.

Quality management should be designed in such a way that processes can directly be influenced as required by the respective results. Manufacturing products that do not correspond to quality regulations and standards should be stopped quickly and efficiently to minimize losses in production as well.

Task 8: Information management

Information has become an essential resource of modern companies. The timely provision of information required for production, such as shop papers and bills of material, have become a basic prerequisite for efficient production. MES information management supports modern production in this respect. In particular, pieces of information are prepared and made available on workplaces or data input stations. Events that affect processes are generated in order to quickly identify erroneous processes and remedy them. In addition, other MES tasks can be triggered by these pieces of information.

An MES solution such as HYDRA has to "fulfill" these duties with respective system functions or software modules (see Fig. 3.4):

PDC	Plant data collection
MDC	Machine data collection
DNC	Administration of machine control programs
SFC	Shop floor control
T&A	Time and attendance
AC	Access control
QDC	Quality data collection
PDP	Process data processing
TRM	Tool and resource management
MPL	Material and production logistics

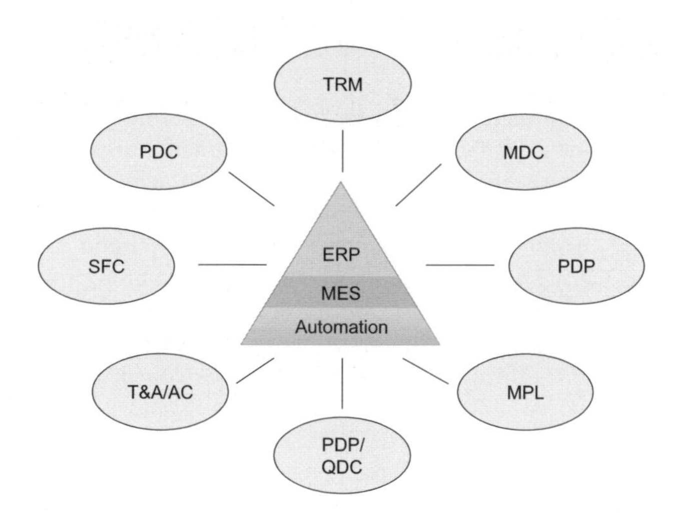

Fig. 3.4 HYDRA software modules representing manufacturing execution system (MES) functions (*PDC* plant data collection, *TRM* tool and resource management, *MDC* machine data collection, *PDP* process data processing, *MPL* material and production logistics, *QDC* quality data collection, *T&A* time and attendance, *AC* access control, *SFC* shop floor control)

3.5 Data Collection During Manufacturing

Data collection is a particularly important task of an MES system because it also constitutes the basis for ERP processing. The possible fields of application of RFID technology almost exclusively affect the scope of collection functions of an MES. The variety of production objects and their automated identification, the ergonomics of screen sequence, plausibility checks of recorded data, and data input technologies are altogether versatile aspects that must be taken into account when planning and realizing an MES solution so that a reliable interface develops and the MES system can fulfill its objectives in the best way. The desire for data collection that does not cause additional effort for the operator within production is something all companies have in common. This may sound utopian, but it needs to be considered thoroughly when the data input stations are being equipped and the collection functions are being designed. Data collection is distinguished by systematic demands to ensure good data quality:

Ergonomics: User-friendly operation is a prerequisite for a successful implementation (see Fig. 3.5).

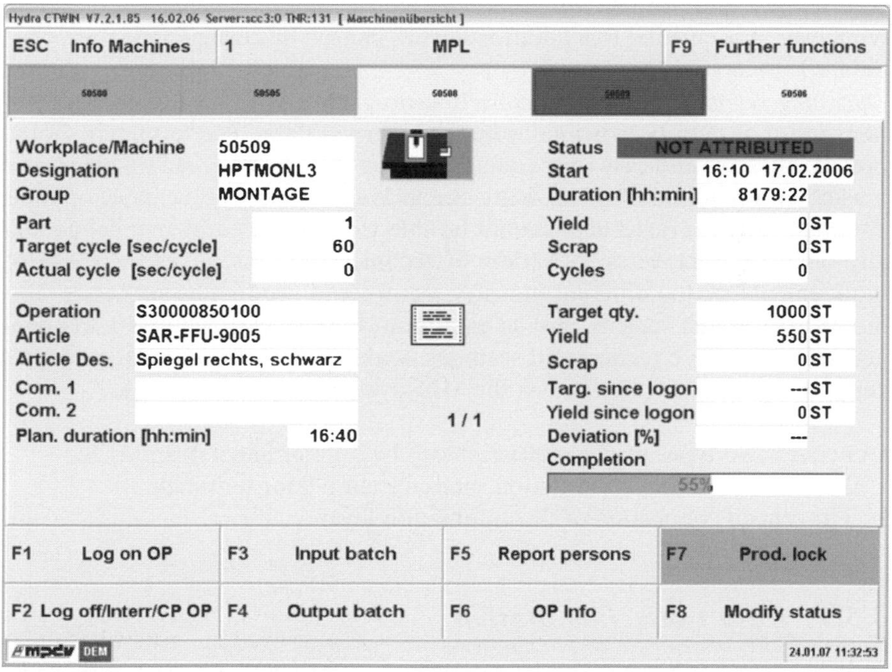

Fig. 3.5 Example of a graphical user interface (via touch screen)

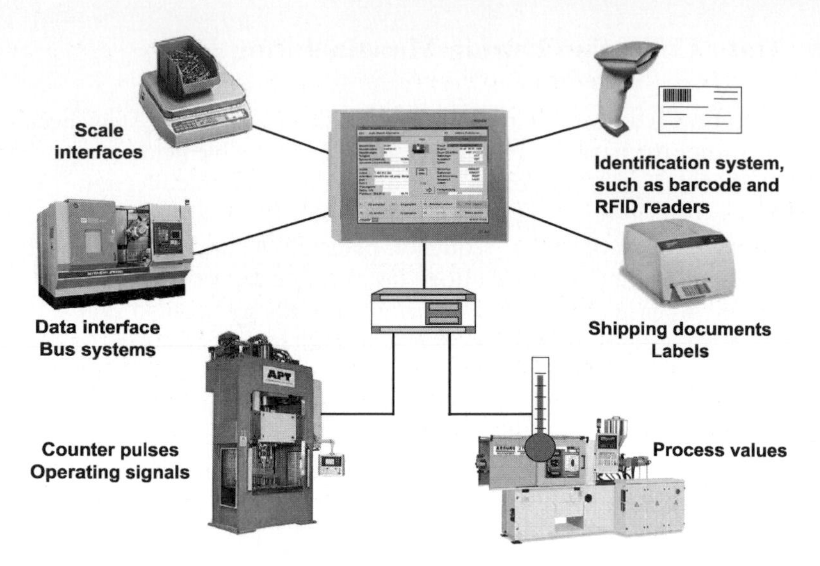

Fig. 3.6 Possible fields of application of an MES HYDRA data collection station

Plausibility and integrity of data: Continuous review of consistency leads to increased process security. Plausibility checks and completeness guarantee high data quality and are therefore a crucial factor for the benefits of collection. Moreover, it is ensured that additional work caused by correcting the recorded data is kept at a minimum.

Operating reliability: The data collection program has to be able to work online, and it should be possible to buffer recorded data. The terminals should provide effective and easy to use data collection dialogs. The data transport and master data required for plausibility checks have to be made available on time. Automatically captured data should be able to be replaced by manual data at any time, and vice versa. A variety of technical options serves the range of requirements of the different industry sectors and production processes. Terminals with touch screens, mobile data input devices with wireless LAN connections, electronic readers and scanners, scales, machines, and plant controls support ergonomic collection via the MES system (refer to Fig. 3.6).

Generally, two types of data collection can be differentiated in an MES:

- The local data collection station such as a shop floor terminal
- The central connection of the automation layer

3.5.1 Data Collection Station

The data collection station of an MES system is no longer only a receiver, as was the case for the double-line data input devices that were used in the early stages of plant data collection and that are still used today. Today's data collection station is also an information center for the operator.

To have information in the right place, rapidly, coherently, and ergonomically—these are the requirements that must be fulfilled by providing information via the MES system in production. The data collection system shows the current status of all production objects and allows retrieval and setting of data and regulations and forwarding of them to the respective automats and machines. The dialogs should be self-explanatory, and the single elements on the display have to be visible and clearly arranged. These demands are to some extent contrary to the style guides generally applied in Microsoft Office automation that demand colorful and sophisticated structures. The operators, however, need to be able to find their way quickly through production without losing time for data collection. For this purpose, the MES has to synchronize the data input equipment with the operators' work flow to guarantee smooth operation within the manufacturing process and ensure data quality at the same time.

3.5.2 Connection of the Automation Layer

Besides local collection, an MES has to be able to connect, for example, complete process control systems to integrate the process steps recorded there as well. The decentralized collection via shop floor terminals (data collection stations) for the staff working on the production level as well as the connection of process control systems such as RFID logistics systems or other subsystems are often required in manufacturing companies.

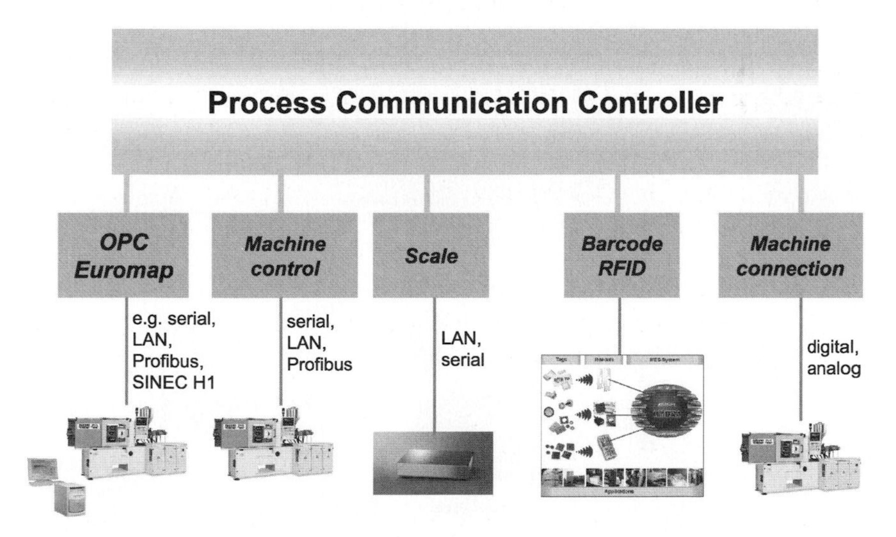

Fig. 3.7 Connection of automation layer with HYDRA-PDC

This can be explained by means of the HYDRA process communication controller (Fig. 3.7). This system module uses a programming interface (API) and the following standard interfaces to integrate any subsystems:

- RFID systems
- OPC (open-link enabling for process control)
- Direct connections to controls such as Siemens S7
- Dry contacts to quantity counting and disturbing signal collection
- Drivers to proprietary field bus systems such as Profibus
- Drivers to machine control networks
- Direct connection to scales or whole scale systems

For these RFID systems in particular, single readers are connected, such as via the RS232 interface.

But if several readers need to be integrated in an RFID system, industrial bus systems such as Profibus DP, Interbus, CAN-Bus, Modbus, WLAN Ethernet, and so on are recommended.

Thus, radio data transmission systems represent an alternative to cabled networking. This alternative often pays off as the costs for cabling are saved. Identification technology is of interest only if it can be integrated into the respective hardware and software of the planning, transport, administration, and warehouse management systems. Transparent and standardized protocols that can also be provided by MES should be available.

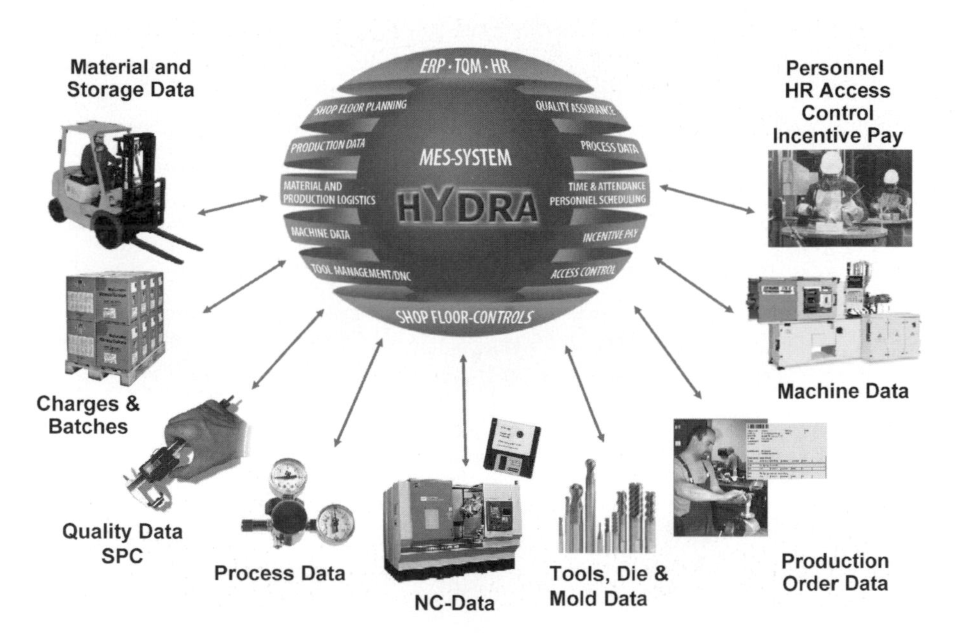

Fig. 3.8 Horizontal integration via the manufacturing execution system HYDRA

3.6 Flow of Information

To clarify the above-mentioned coherences, the flow of information from the automation layer to the company management level will now be presented in a simplified way using an example based on SAP software and HYDRA.

The task of MES HYDRA is to record data on the production level. In this context, the total scope of accumulating data must be able to be processed via the most diversified technologies, as shown in Fig. 3.8. Furthermore, these data must be able to be checked for plausibility and be aggregated at the time of recording.

The processing specifications and target values needed for the operations are delivered by the ERP system. The data required by the MES system are transferred to the applications of the SAP system in a demand-oriented (timely) way after having been aggregated and checked for plausibility. To guarantee the cross-system process capability of the overall solution, the functions of both systems synchronize via the SAP integration technology, which is realized with interfaces that cannot be "perceived" by the users.

Figures 3.9 and 3.10 present some SAP applications that are relevant for production, showing the necessary information flow from and to the manufacturing layer.

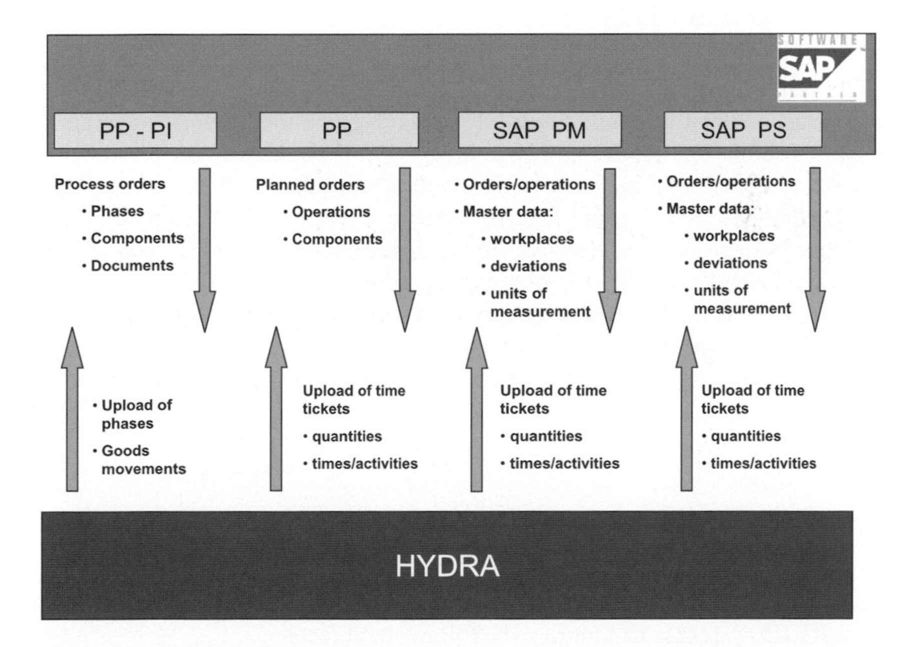

Fig. 3.9 Example: flow of information from and to SAP applications and HYDRA

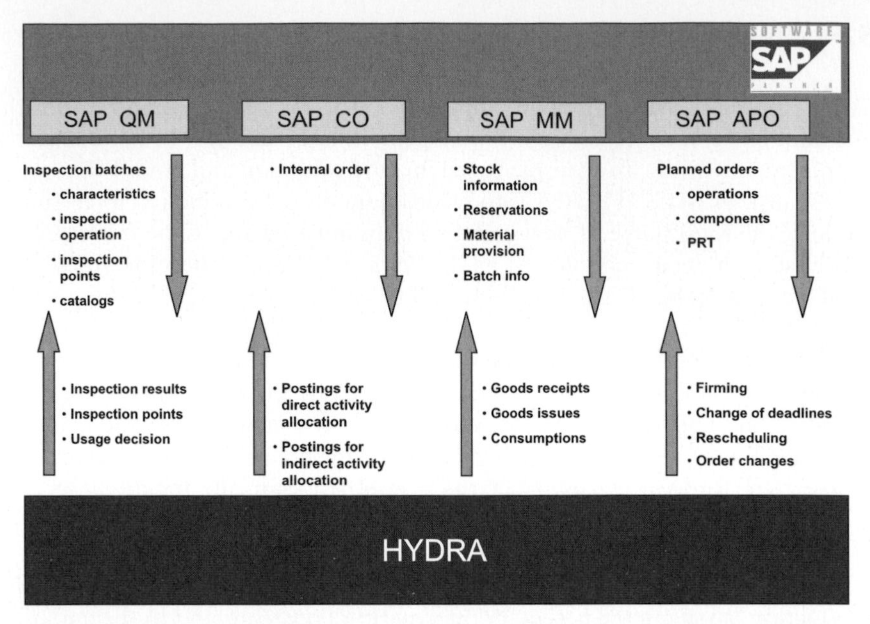

Fig. 3.10 Example: flow of information from and to SAP applications and HYDRA

3.7 RFID and MES

RFID systems can be used in nearly all functional areas of an MES. In general, organizations must assess individually which type of identification system will be most useful. The RFID chip will always compete with other identification systems, and, unfortunately, this book cannot provide the reader with a patent remedy that says which identification system best suits which area of an MES solution. Nevertheless, a decisive factor is the fact that RFID technology is only useful as an integral part of the previously described MES functions and then becomes an individual application by means of which production processes can be supported and optimized. To underline this situation, the possible fields of RFID application are presented in different examples in the following sections.

3.7.1 Identifying Objects in Production

Production is based on the interaction of a series of objects. The primary task of an RFID system is to identify the following objects.

3.7.1.1 Order/Operation

Orders are generated in the ERP environment and are usually resolved into operations via bills of materials. At the same time, they are used as cost collectors to receive the services rendered and the quantities produced. Normally, the order or operation is seen as a central key for many expenses and information accumulating within production. According to this organizational particularity, an MES should use the order as a central key for information, too. In process industry, the order shares its role with article-based orders that are divided into batches. In this case, batches have to be identified individually.

3.7.1.2 Material

Materials are kept as input materials of processes and also as intermediate and finished products within ERP and thus in the MES system as well. Because it is typical for ERP systems to care only for the yield and the scrap rendered for the materials balance, it is the duty of an MES solution to regard all material types occurring within production, namely yield and scrap, as well as rework. In ERP, stock is often only considered to be material that is located in the warehouse, but a large amount of WIP, which has to be managed and transported in the best possible way, is used in the production. For this reason, an MES should manage inventory in a detailed way and based on the production cycle in intermediate storage, which is commonly called WIP materials.

3.7.1.3 Resources and Production Resources and Tools

Resources required for the production process and which are available only in a limited capacity at the same time are planned, assigned, and occupied by the production process. Tools, especially trained staff such as fitters or employees in charge of quality, as well as special handling devices and required equipment belong to these resources. (See Fig. 3.11.) In many production environments, the resource tool is more important than the machine. Maintenance related to tools is based on the collection of quantities and times rendered. Due to maintenance and repair, resources are available only in a restricted way. To avoid errors in planning and materials requirement, it is essential to keep the status of an operating resource as exact as possible.

Due to completely different surrounding conditions (temperature, fouling) and the different properties and conditions of objects, it is very important to be able to use a broad range of flexible identification systems.

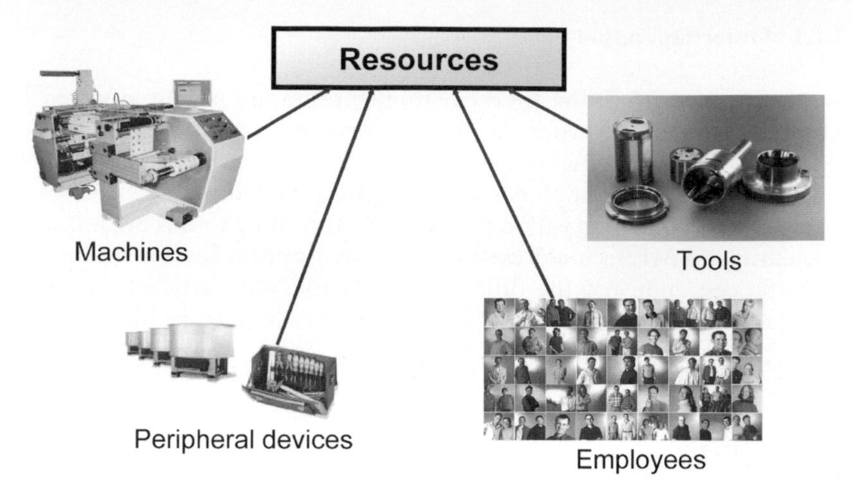

Fig. 3.11 Resources for the production process

3.7.2 Realization of Online Interfaces

As already described in Sect. 3.5, an important MES function is integrating the different IT systems of production (process control systems, insular solutions to control aggregates and machines, machine controls, etc.) into the ERP level. Therefore, information exchange among the IT systems involved is required. Systems are connected technically via networks (e.g., Ethernet or field bus system) or direct connections (e.g., RS232 interface). In the future, a standardization of the data format on the RFID chip might be used to render data exchange more flexible—between the different IT systems in general and to process control systems or subsystems for controlling or automating machines in particular.

Information frequently has to be transferred from one system to the other. The method commonly used today is to provide the transport units or products with a barcode as identification. Information is transferred from one system to the next via an IT-specific connection of both systems. Besides the actual interface programming, a more or less complex IT infrastructure is required as well. The IT infrastructure increases costs if the interface must be online because the logistic process requires it. The pieces of information must be available when the product or material to be identified arrives at the other system.

When RFID chips are used for the data export, the savings potential lies in the IT infrastructure (cabling, connection of systems via network, additionally

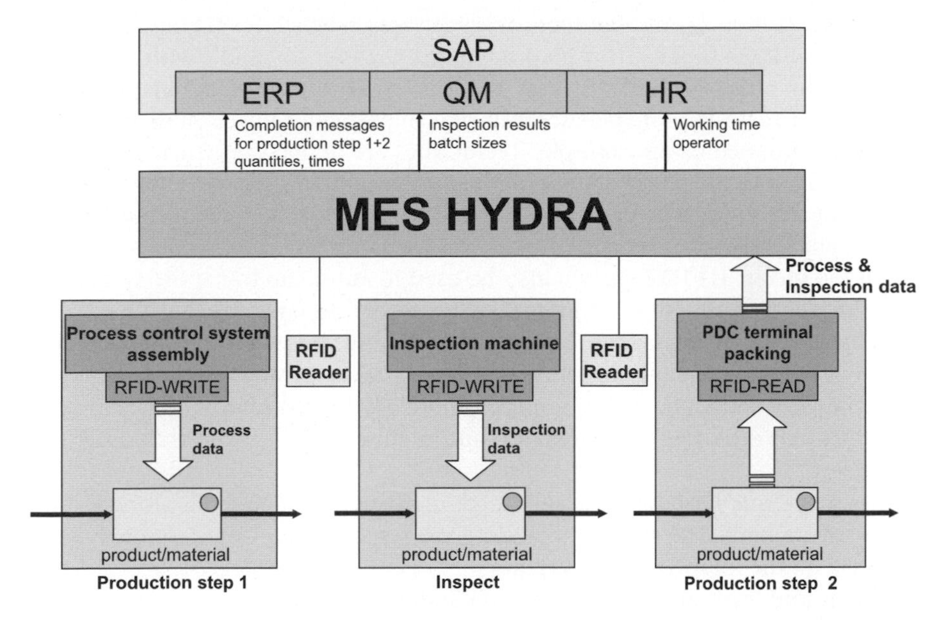

Fig. 3.12 Example: online interfaces via radio frequency identification (RFID)

required interface resources, or even interface processors). The interface programming is certainly still required, but it could be simplified by standardizing the contents of RFID tags as already done in other areas (Transmission Control Protocol/Internet Protocol, OPC, etc.). The synchronization of two systems exchanging information via RFID is virtually given automatically, as the material flow (RFID tag on the product) and information flow (RFID tag that accompanies the product from system to system) are synchronized (Fig. 3.12).

3.7.3 Intra-Enterprise Logistics

In the future, the RFID code carrier will play a decisive role in global supply chain management. In this area, the cost and benefit analysis will definitely be positive as the RFID tag is used to a large extent within a closed loop. Consequently, the tags do not get lost, for example, when goods are delivered to the clients, but they circulate within the company's production. This may certainly change in the global supply chain when customers and suppliers are included. Then, the RFID chip is frequently used to label the products and the goods delivered and to refer to these pieces of information in the advanced supply chain.

It goes without saying that there are also cases in which RFID tags are "built into" the product because this method might be easier to handle with respect to production, although RFID is only used internally for transport management and flow of material. The plastics industry, in particular injection die molding, can be mentioned as an example. Techniques are applied in which the RFID chip is added to the plastics mass of the injection molding part. Thus, the chip is within the product when it is delivered to the customer or during further processing (assembly).

In the future, RFID tags will also be used to automate transport systems in intralogistics, i.e., to control unattended transport systems.

This applies particularly to processes as they are known from random storage. For example, the market already offers systems in which individual goods are recorded automatically or semiautomatically in real time. The single storage locations are saved and identified automatically. The slogan is finding, not searching.

Such systems have been designed for clearance surface, bulk storage, and high-bay racking. The system, including the respective hardware and software, is based on the following components:

- In-ground location transponders to identify location
- Antenna mounted under the forklift to pick up the location transponder
- Transponders mounted onto the flexible part of the telescope to record height (height transponder)
- Antenna mounted onto the rigid part of the telescope to pick up the height transponders
- An antenna installed in between the forks to collect the circular transponders
- Alternatively, a barcode scanner aiming at product identification

3.7.4 Quality Management

When it comes to quality management, the inspection of material or components, either manually or by means of inspection machines where inspection results are saved on an RFID chip, is considered first. Basically, this is reasonable only when it is impossible or difficult, due to IT reasons, to save information online on a central database of certain inspection equipment. Otherwise, along with saving the inspection data on the RFID chip, the information also has to be archived. In some areas of the supply chain, such as in customer/supplier relations, the intent is to transmit the inspection values themselves, thus transferring the quality results from the supplier to the customer, a practice that could be extended up to an automatic (paperless) transfer of test certificates. However, this would require standardization of data formats. To what extent an RFID technology can actually replace the networking of companies, also in view of electronic data interchange (EDI) and whether redundant communication channels will be created that, in the end, have only a limited savings impact, should be surveyed very closely.

The identification of "materials," "components," and "tools" is contrary to this. Within the production process, it is advantageous to lock objects by labeling an RFID chip that is read during the processing of the object every time it comes to plausibility checks by an MES, whenever materials are processed within a production step, or whenever tools are implemented before a production step is started. This certainly also applies to process parameters that are possibly required or need to be read in the next processing step. Within a production where the automation components are already widely linked via an MES, the benefits of transporting information via an RFID tag are basically limited. But here, too, the RFID tag provides the opportunity to integrate such pieces of information into the process cycle basically offline and by manual action.

Example: A visual inspection of parts, which has not been planned in advance, takes place in the warehouse. An employee detects that a component is damaged and cannot be processed in the required quality in the next stage of production. Whereas employees would normally need system access (data input device, screen, or similar), they could now just put a lock flag on the available RFID tag. A simple, portable writer/reader would suffice in this case. The employees can be sure that the RFID tag is checked within a subsequent processing step and that the information is provided at the right place and at the right time in order to prevent the faulty material from being processed. If the inspection takes place on a predefined workplace that is already equipped with system access, then this method is not very advantageous.

3.7.5 Access and Attendance Control

Noncontact code carriers such as the RFID tag or transponders are widespread in applications for time and attendance and access control.

Authorized employees receive a personalized transponder, such as in the form of a credit card or key ring. Access is granted or denied at doors, entries, turnstiles, and so on by respective readers. Corresponding MES systems establish connections to available pictures, access days and times, time recording, and flextime accounts.

In production areas with high security standards, such as pharmaceutical and medical engineering, these code carriers are increasingly used for automatic identification of employees. Within the production cycle, legitimation is important not only for access control to the MES or ERP system but also to document the production steps. In these production areas, a system complying with the FDA, for example—in which a "second set of eyes" is required—specifies that employees entering data into the system have to be legitimated. By using RFID in this context, operation cycles can considerably be reduced compared to recording barcode badges manually.

3.7.6 Shop Floor Control

Regarding the usability of RFID, intralogistics and shop floor control are closely linked. The required MES functions overlap.

As already described in the intralogistics chapter, the RFID chip allows to automatically transport process values relevant for production from one production step to the next. This means that, for example, recorded pressures or temperatures are used as input parameters for production in a next production step. Moreover, these options can also be used to partially automate production cycles by deciding automatically, because of saved quality data, whether automatic rework processes have to be "included".

In the automotive industry, many production cycles are performed automatically or semiautomatically. For instance, automotive manufacturers and suppliers worldwide equip the body shell with a transponder that stays with the vehicle during the whole production process and is also controlled by it (the addition of components, colors, engine variants, and so on) at a very early stage of production. The transponder code can later be used to identify the vehicle in the service points in cases of maintenance or repair in order to procure spare parts (e.g., information on the assembly group version).

Besides the direct labeling of the vehicle, the transport frames, or skids, are provided with a transponder because these skids must endure a temperature of 220°C when the lacquer is annealed.

When adding modules of different vehicle types onto production lines within random production, a unique identification of the components and assembly groups supplied (cockpits, engines, seats, etc.) is indispensable. It is not only important to mark the modules to provide for a correct assignment, but the mounting devices also have to be characterized with transponders for this purpose.

However, in this context it is also a determining factor for the MES or ERP system to provide the necessary logic in order to be in the position to use RFID data for automated shop floor control.

Looking into the future, it could be imagined that MES functions—without the context of the total system—will be provided locally on workshop facilities and will be able to be linked via RFID data in terms of a service-oriented architecture that is provided by an MES system and that requires a uniform definition of data structures on the RFID chip and modes of operation within the service. To realize such logic approaches, researchers have intensely worked on what was commonly known as "agent" technology in the past. Through this connection, research aims to establish an "intelligent overall logic" from individual logic approaches that are actually independent of each other.

Finally, it can be concluded, as the scenarios described above show, that the integration of RFID technology into MES functions constitutes actual benefit for users in most applications within manufacturing.

3.8 Summary

In this chapter we gave an overview of an MES and its role within a manufacturer's IT infrastructure. We outlined the typical structure of a manufacturing company along a layered model of the company's processes and supporting IT functionality. Against the background of this model, we presented the functionality of an MES and described how it is integrated into the enterprise information infrastructure. We discussed tasks and functions of an MES in detail, as well as the relevant processes for an MES. We set a special focus on data collection during manufacturing, which covers technologies for data capturing as well as data processing and communication on lower system levels. Here we explicitly addressed integration issues with devices on the plant floor. The chapter ended with a discussion of RFID's impact on various functional areas of an MES, including why these functional areas are influenced by RFID technology.

Chapter 4
Six Case Studies

Coauthored by Lenka Ivantysynova and Holger Ziekow

In this chapter we present six case studies illustrating the use and potential of RFID in the manufacturing industry. Each case study focuses on the production processes in a specific plant. The participating companies are from the following industries:

1. Automotive industry: manufacturer of airbags (AIR)
2. Automotive industry: manufacturer of sliding clutches (CLU)
3. Automotive industry: manufacturer of engine-cooling modules (COO)
4. Steel and mill industry: manufacturer of cast parts (CAS)
5. Electronics industry: manufacturer of connectors (CON)
6. Packaging industry: manufacturer of packaging (PAC)

The majority of the companies already use an SAP enterprise resource planning (ERP) system or plan to do so. Most of them also use an MES. Of the analyzed companies, COO is the only one already using RFID in production. We found that two reasons dominated the companies' interest in RFID: On the one hand, they expect their customers to demand RFID solutions in the future, and they want to be prepared for this. On the other hand, because of specific customers' demands and the need to obtain competitive advantages, the companies aim to improve the tracking of their production processes.

All six companies assumed that RFID could either be a solution for cases in which barcode technology is not applicable or be an advantageous alternative to barcode technology. The case studies were designed to verify this hypothesis.

4.1 AIR: Airbag Manufacturing

The investigated plant of the airbag manufacturer, AIR, assembles complete airbag modules (Fig. 4.1) and produces airbag covers. Currently, AIR is using

O. Günther, W. Kletti, U. Kubach, *RFID in Manufacturing*
DOI: 10.1007/978-3-540-76454-0, © Springer 2008

Fig. 4.1 An airbag in action

barcode technology for tracking and tracing all materials during the whole production process.

However, AIR needs to decide whether to continue and possibly expand its barcode technology or switch to RFID. The objective of the case study was to evaluate whether there could be benefits of using RFID at a specific production plant.

4.1.1 Current Situation

For AIR, it is important to ensure and document that all produced items have passed each manufacturing step. There are three reasons for this:

- First, an increasing number of customers require detailed reports about their orders. Some customers want to track every step of the production process in order to facilitate the processing of future warranty claims and to narrow recalls.
- Second, AIR needs to be able to protect itself against product liability claims. For instance, consider a car accident in which the airbag did not open. AIR must be able to ensure that the airbag was produced correctly and that the error lay somewhere else (for instance, in the electronics). If AIR cannot prove proper assembly of the airbag, it would have to take full responsibility, pay the direct costs, and recall a large number of airbags.
- Third, there is a need for plausibility checks during the manufacturing process in order to prevent errors.

Today AIR uses barcode technology for ensuring and documenting that items have passed each manufacturing step. However, AIR is about to change its production processes and scan items' barcodes after each production process has been completed. At this point (in 2007), AIR has already changed half of its production processes. In the remainder of this chapter, we consider the situation *including* this new barcode infrastructure.

A barcode is attached to every cover at the very beginning of production. It is used as a database key that (1) allows recording when the cover passes a specific production step and (2) enables automatic plausibility checks in the back-end system. During production, numerous scan transactions are executed. The company conducts manual as well as automatic scan processes, and manual scanning is time-consuming. Especially when the covers' barcodes have to be scanned, it is also cumbersome to ensure a line of sight between reader and barcode. The covers must be turned upside down to enable scanning because barcodes are often applied to the inside of the cover. Also, scanning may fail and have to be repeated, resulting in additional delays. AIR actually analyzed how time-consuming manual scans can be and found that a scan takes 4 seconds on average.

An airbag consists of three main parts: cover, cushion, and inflator. At the investigated plant, only the production of the cover and the final assembly of the airbag take place. All other parts are bought from other AIR plants.

The production of an airbag can be divided into five steps: *injection molding*, *flash removal*, *special surface treatment, varnishing*, and *assembly* (see Fig. 4.2).

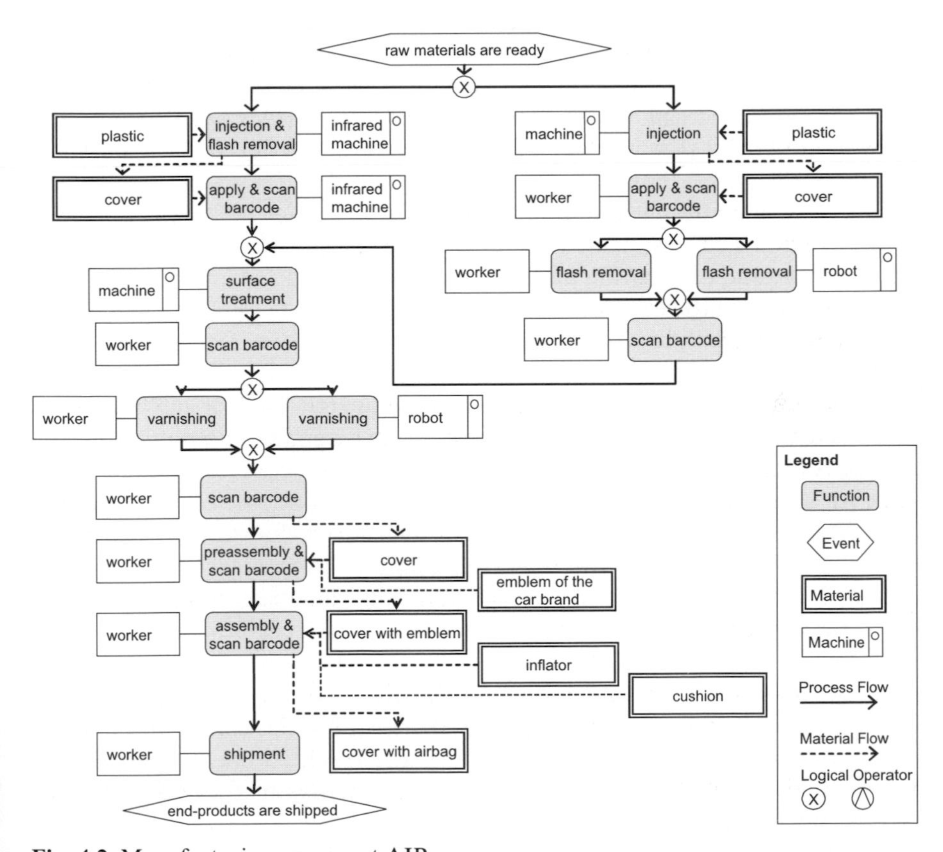

Fig. 4.2 Manufacturing process at AIR

The production starts with the injection of the cover. Then excess plastic is removed in the flash removal step. Next comes a special surface treatment, which increases the quality of the varnishing. This step is not always performed because some customers do not explicitly permit the treatment. Subsequently, the covers are sent for varnishing. After this step, the airbag is assembled; that is, covers, cushions, and inflators are mounted together.

In the following we describe each production step in more detail. In the first step of the production (injection molding), plastic is injected into a mold and formed to a cover. Then a barcode is applied to the cover of the future airbag, and the barcode is scanned. As mentioned above, the barcode is used in the following production processes for tracing the product.

The second step is flash removal. Excess plastic at the parting lines of the mold is removed in this step. There are three different methods of flash removal: manual, with special robots, or with the use of an infrared machine. The method used depends on the type of cover and the size of the order. The infrared machine is used only when the revenues expected from the order allow writing of a new program for the infrared machine. If this is not the case, the robots will be used for flash removal. Manual flash removal is used only in cases of small orders, when the cost of changing the program for the robots also does not pay off. With both robots and manual removal, workers have to manually scan the cover's barcode after flash removal. If the infrared machine is used, the first two steps, injection molding and flash removal, are executed together, and only one automatic scan of the barcode is conducted. Here, plastic is injected into the mold, and then the excess parts are removed. After that, a barcode is automatically attached to the inner part of the cover and is scanned.

The next step after flash removal is the special surface treatment, a special activation of the cover's surface. In this step the product is exposed to high tension with the aim of ensuring a better lacquer quality. The barcode is scanned after the special surface treatment is completed.

In the following varnishing step, the covers are varnished in machine complexes or a manual varnishing facility. Varnishing a cover requires temperatures of up to 80°C. After the covers are varnished, the barcodes of the covers are scanned again. Covers leave and enter the varnishing facility on a cart if varnishing is done manually.

After varnishing, a car logo is mounted to the front side of the cover. In Fig. 4.2, this is depicted as preassembly. Then in the fourth step, called assembly, the cover, cushion, and inflator are mounted together. Before they are mounted, barcodes of all three parts are scanned. These scan transactions trigger a plausibility check in the back-end database, which ensures that the correct parts will be mounted together. Workers must wait for the response from the database before they can carry on with their task. According to the plant's IT managers, about half a second is a tolerable response time, which they try to always ensure. Depending on the number of scan transactions, however, the response time of the back-end system may be higher, delaying the manufacturing process accordingly.

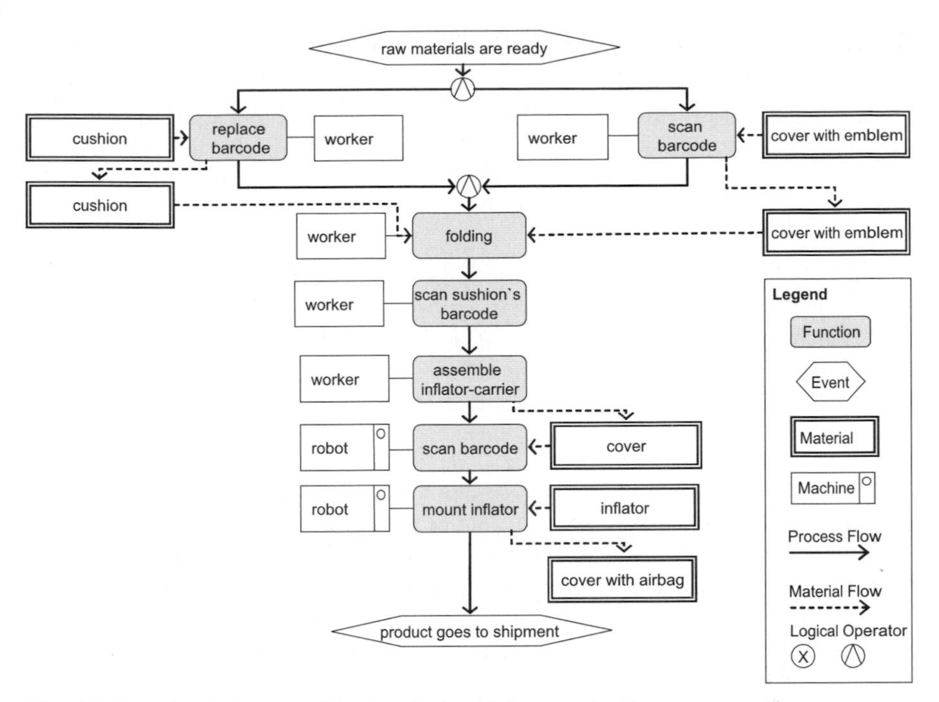

Fig. 4.3 Details of the assembly step during airbag production

The assembly itself is subdivided into several steps and may vary depending on the airbag system being produced. We discuss assembly using an example with five different machines and two workers; details of this process are shown in Fig. 4.3. While one of the two workers changes the place of the cushion's barcode, the second takes a new cover and scans the cover's barcode. Subsequently, the second worker takes a cushion, folds it into the cover, and scans the cushion's barcode. After this, the inflator carrier is mounted to the cover. Then the first worker places this semifinished product, which consists of a cover, a cushion, and an inflator carrier, into a machine. The machine assembles this semifinished product with the inflator. As the semifinished product and the inflator are put into the machine, they are automatically scanned for a second plausibility check. With the assembly of this semifinished product with the inflator, the assembly step is completed.

After assembly it is no longer possible to scan the barcodes on the cover or cushion because they are hidden inside the product. However, the inflator's barcode can be seen at the back of the completed airbag. Further use of this barcode is not feasible due to different customer requirements for barcode labels. Therefore, new barcode labels are applied according to the customer's restrictions and requirements. Finally, airbags are boxed and sent to a warehouse (shipment step in Fig. 4.2). The airbags that are put into a container are registered accordingly in the back-end system.

4.1.2 RFID Perspectives

AIR currently does not use RFID in manufacturing. Data management and tracking of products are realized exclusively with barcodes. In this section, we discuss possible RFID applications and their potential benefits for AIR's production process. We address the following five areas of RFID potential at AIR:

1. Reducing communication with the back-end system
2. Reducing manual scan transactions in production
3. Reducing manual scan transactions in the warehouse
4. Avoiding barcode print quality issues and reducing penalties
5. Avoiding customer-specific barcode printing

4.1.2.1 RFID Application Scenario #1: Reducing Communication with the Back-End System

Every scan transaction of a barcode triggers a query in the back-end system to get the required information about the product. The production process must wait for the answer to the query to ensure the correctness of the production flow. According to AIR's IT management, a response time of up to half a second is tolerable. Considering that in the current production process not all planned scan transactions are executed yet, the back-end system will be burdened further when all processes include a scan of the items' barcodes. As a result, the response time may exceed the tolerable latency.

With RFID adoption, the problem of time-consuming back-end transactions could be overcome: The covers could be tagged with writable RFID tags. Then the necessary information for the production process could be stored on the tag. As the cover passes through the production process, RFID readers could communicate with the tag on the cover and read the required information. Synchronous communication with the back-end system would not be necessary anymore if all information were saved on the RFID chip.

An important point to consider is that RFID tags can store different amounts of data depending on the type of tag. The company would need to analyze which information may be required at which stage of the production process. The second issue to be decided is which part of the required information should be stored on the tag. It may not be possible to store all information required during the production process. Thus, it must be decided which information may have a stronger effect on reducing the response time. Based on this optimal amount of information, it would be possible to calculate the required memory capacity.

4.1.2.2 RFID Application Scenario #2: Reducing Manual Scan Transactions in Production

AIR has already partially implemented scans of the products' barcodes after some production steps. Many of these scan transactions are conducted manually and account for a significant percentage of the employees' activities. Applying RFID may allow automation of many of these scan processes and help save time in production.

As mentioned above, the barcodes are applied to the inside of a cover. Therefore, the cover must be turned upside-down before the barcode can be scanned. This takes extra time, about 4 seconds on average. When the number of scan processes is considered, this leads to a significant loss of time and money. Assuming that a worker costs 25 euros per hour, i.e., about 0.7 cents per second, the extra cost is 2.8 cents. Table 4.1 shows how much scanning time is spent in total during the whole production process. We assumed that the execution time of scan processes done automatically by machines is zero. Scanning by machines does not cost any extra time because the whole production process is not delayed. When workers scan, the production gets delayed because they do not operate the machines while scanning the barcodes.

Table 4.1 represents two different options regarding scan transactions in the production processes. The first option starts with injection molding and flash removal. Barcodes are applied and scanned automatically. Scans are not

Table 4.1 Time cost of scan transactions during different production steps

Option 1	Scanning	Time in seconds	Option 2	Scanning	Time in seconds
Injection and flash removal (infrared)	Automatic	0	Injection	Manual	4
			Flash removal	Manual	4
Special surface treatment (optional)	Manual	4	Special surface treatment (optional)	Manual	4
Varnishing (manual or automatic)	Manual	4	Varnishing (manual or automatic)	Manual	4
Preassembly	Manual	4	Preassembly	Manual	4
Assembly	Manual (cover)	4	Assembly	Manual (cover)	4
	Automatic (cover and cushion)	0		Automatic (cover and cushion)	0
	Automatic (inflator)	0		Automatic (inflator)	0
Total scanning time		16	Total scanning time		24

executed until both the injection molding and flash removal steps are completed. Therefore, these transactions do not need any time. After the special surface treatment, varnishing, and preassembly steps, barcode scans are executed manually. Executing the varnishing step manually or automatically does not make a difference in terms of scan transactions because after both cases, barcodes are scanned manually. In assembly there are three scan transactions, one of which is executed manually. Other scan transactions in assembly step are executed automatically and therefore hold no further potential for time savings with an RFID solution.

The second option in Table 4.1 starts with the injection molding step. In this option, flash removal and injection molding are executed separately. The barcode is applied and scanned right after the injection molding step. The scan transaction is conducted manually. Another manual scan transaction is executed after flash removal. Special surface treatment, varnishing, and preassembly steps follow.

After each of these three steps, a manual scan transaction is conducted. The last step of this option is assembly. It contains three scan transactions, but only the cover is scanned manually; other scan transactions are executed automatically in the machine.

The difference between these options lies in the first two steps. If injection and flash removal are conducted in the infrared machine together, as in the first option, barcodes are scanned automatically. If the second option is chosen, the barcode is applied right after injection molding and is scanned manually after both injection molding and flash removal. As a result, the second option takes 8 seconds longer than the first option.

It would be possible to design the RFID adoption in such a way that reading RFID tags would be executed simultaneously to another transaction that must be carried out in any case. For instance, at the automatic varnishing step, the RFID reader could be installed in such a way that it would read RFID tags while the cover is sent out of the machine. Thus, it would be possible to gain more than 16 seconds for the first option and 24 sesonds for the second option, resulting in a savings of approx. 11 cents in the first and 17 cents in the second case.

When varnishing is conducted manually, the covers could be read in bulk. The facility is in a room with a door at both ends. After the covers are varnished, they are placed in a metal cart. When the cart is full, a worker pushes it out of the varnishing facility. After this, all covers in the cart must be scanned manually.

As manually varnished covers pass through the varnishing facility door, RFID readers could register the covers. However, there are some special challenges in using RFID in this facility: The carts and the door are made of metal, which may cause shielding of radio signals. Several means can be considered to deal with the metal environment. First, the best positioning of reader antennas should be investigated. The shelves in the carts build a stack of several horizontal metal layers; consequently, vertical radio signals are likely to be blocked.

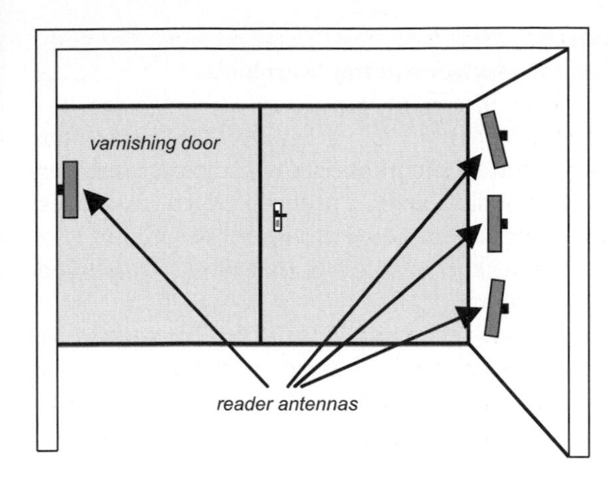

Fig. 4.4 RFID reader installation on the door of the varnishing facility

Therefore, we suggest installing antennas on the side of the varnishing door to enable signal spread between the layers (see Fig. 4.4).

Another possibility is to replace the metal carts with carts that do not negatively affect radio frequency communication. However, the cost of replacing carts must be considered. The plant has 140 metal wagons used for transporting manually varnished covers. The cost of a new carts lies between 200 euros and 300 euros.

However, the door of the varnishing facility is also made of metal and can therefore still cause disturbing reflections. This means that changing the carts probably would not be enough, and other measures must be taken into account to inhibit reflections from the metal door. Therefore, the use of RFID tags, which are less influenced by metals, must be considered. That is, frequencies from the UHF spectrum are favorable because these are known to be more resistant to disturbances. Yet only a field study could fully clarify whether the varnishing facility would allow the use of RFID.

Besides using RFID tags on the cover, AIR could consider using them on the other parts of the airbag as well. In the assembly step, cushions and inflators are involved, and their barcodes need to be scanned, too. By tagging cushions and inflators with RFID chips, even more time could be saved. But the additional cost for the RFID chips must definitely be taken into account. Another point to consider is the question of which tag type should be used for covers, cushions, and inflators. Because all of these tags remain on the product and can be used only once, it seems most reasonable to use low-cost tags for cushions and inflators. This would decrease the overall RFID adoption costs. However, using different types of RFID tags for cushions, inflators, and covers must be intensively researched and the benefits of different combinations taken into account.

4.1.2.3 RFID Application Scenario #3: Reducing Manual Scan Transactions in the Warehouse

In the shipment step, the end products are placed in containers and sent to the warehouse. At this stage, management of end products is no longer handled at the item level but rather at the container level. Products, which are placed in containers, are booked to the containers' account. Booking airbags to a container or booking out an erroneous entry are tasks that are accomplished manually and therefore are time consuming.

RFID readers and tags could provide more reliable and flexible warehouse and delivery management. RFID readers could be installed at the warehouse door so the accuracy of a container's contents could be checked automatically. By employing this solution, it would be possible to decrease the time-cost of scanning containers and of failures due to human intervention.

Realizing this scenario would also allow AIR's customers to use RFID at their inbound logistics, because the arriving products and containers would already be equipped with RFID tags by AIR. If the customers make use of these tags, cost-sharing models might be feasible. Reusing the existing RFID tags at the customer's side can be perceived as a service of AIR and thereby be a competitive advantage.

4.1.2.4 RFID Application Scenario #4: Avoiding Barcode Print Quality Issues and Reducing Penalties

After production is completed, new barcodes are applied to the assembled airbags and formatted according to customer requirements. The barcodes are used for identification and for ensuring information flow between supplier and customer. Barcodes on airbags must be flawless and correctly applied. Scratches or dirt can make the barcode unreadable. Achieving the printing quality required by the customer poses an additional challenge. The barcode scanners installed at the customers' sites may have problems reading low-quality barcodes. Because many of AIR's customers implement a just-in-sequence (JIS) delivery, each unreadable barcode potentially causes a delay of the customer's production. In such cases AIR has to pay high penalties. To avoid such penalties, AIR puts great effort into printing high-quality barcode labels.

Because barcodes are line-of-sight technology, they must be applied to the outside of the product. If the respective customer would agree to apply RFID tags instead of barcodes, the possibility of label damage would decrease because RFID tags are less affected by dirt and scratches on the label. Furthermore, the difficulties in printing high-quality barcode labels could be avoided. In this case, readability is not influenced by the writing process. The burden of creating the labels is thereby eased, and penalties can be avoided.

The customer (typically an automotive original equipment manufacturer, or OEM) could use the RFID tags that are applied to the cover at the beginning of the production process (see application scenarios #1 and #2). With the use of RFID, the number of airbags with unreadable barcodes and the amount of forfeits could be reduced. More generally speaking, the concrete interest of AIR's customers in RFID is of crucial importance regarding realization of this particular RFID potential.

4.1.2.5 RFID Application Scenario #5: Avoiding Customer-Specific Barcode Printing

Another advantage of using RFID tags would be the potential reductions in effort and costs for printing a wide range of customer-specific labels. Today, a label is applied to the shipment containers and to the finished products according to the customers' demands for information and label formats. If the customer agreed on using RFID, customer-specific information could be written on the tags without much effort. This would reduce the total number of necessary label types, and the RFID tags applied in the production process could be reused for shipping.

Another advantage that can be gained from using RFID tags is the reduction of the cost of labels that the customer demands. Currently, a customer-specific label is applied to the shipment containers and to the finished products. These labels hold information required at the customer side. If the customer agrees to use RFID, the customer-specific information could be written on the tag. This would reduce the total number of necessary labels, as the RFID tags applied in the production process (scenarios #1 and #2) could be reused.

4.1.3 Costs and Benefits

The main reason for considering RFID at AIR is to reduce manual barcode scan processes. Manual scanning accounts for a significant portion of employees' time. AIR's IT staff estimated that the application of RFID would save up to 4 seconds at each manual scan point. Applied to all production lines, this would add up to several tens of thousands of hours of work per year. Additional time could be saved in the outbound shipping processes, where the checkout could be automated. Yet only a few cents could be saved per airbag. Because in this scenario RFID tags remain on the product, the tag cost must be gauged against the savings per airbag. At a price of 20–30 cents for low-cost passive RFID labels, this may or may not be cost effective. Thus, cost-sharing models with subsequent players in the value chain (who might reuse the tags) should be considered.

Another reason for RFID adoption is to reduce the load on the back-end system. Currently, the infrastructure at AIR is at its limits for processing plausibility checks in the back-end system. The planned expansion of the number of scan points would therefore require additional investments in the server infrastructure. If tags with extended writable memory are used, consistency checks could be processed independently of the back-end system. Here, the cost for additional servers must be compared to the running costs for RFID tags with sufficient memory as well as the effort for implementing plausibility checks on local programmable logic controllers (PLCs) or terminal PCs.

Further savings may result from simplifying label management and avoiding penalties for unreadable barcodes. AIR now pays a few thousand euros per barcode when the customer cannot read it. Such penalties could be avoided if RFID were applied. However, this requires that customers agree to switch from barcodes to RFID-based solutions.

4.1.4 Summary

In this case study we discussed the potential benefits of using RFID in AIR's production processes. AIR assembles complete airbag modules and produces covers, which are one of the three parts in an airbag module. We analyzed all production processes in order to define the potentials of RFID adoption at the investigated plant. We examined whether the use of RFID in this production process could be advantageous compared with the current application of barcodes. We outlined five potential benefits RFID may hold for AIR: (1) reducing communication with the back-end system, (2) reducing manual scan transactions in the production process, (3) reducing manual scan transactions in the warehouse, (4) avoiding barcode print quality issues and reducing penalties, and (5) avoiding customer-specific barcode printing. All advantages were analyzed with respect to their resulting costs, including costs for RFID readers, tags, and required software.

4.2 CLU: Sliding Clutches

The investigated plant of a manufacturer of sliding clutches, CLU, assembles sliding clutches for drive trains. CLU constantly desires to improve its tracking process during production. This is because fine-grained tracking data can help narrow product recalls, which account for significant costs at CLU. To capture further tracking data, CLU has started to consider RFID introduction from the material intake on. With the following analysis, we follow up on a previous study that was conducted by a management consulting firm.

4.2.1 Current Situation

At the investigated plant, CLU produces sliding clutches, which consist of several gear wheels and metal rings (see Fig. 4.5). Some parts of these are bought-in and directly assembled. Others parts need further operations in the plant. All of these materials are received at a single intake, which is also used as the shipping area.

The process at CLU usually consists of seven operations that are highly automated by machines: *broaching*, *milling*, *cleaning*, *carburizing*, *hardening* (combined with *grit blasting*), and *assembly*. An abstract model of this production process is shown in Fig. 4.6. Human intervention is limited to loading and unloading, setting up, and maintaining machines, as well as recording production data. Workers interact with the IT back-end system via terminals on the plant floor. These terminals are controlled by the MES HYDRA. The MES is in charge of supervising all operations on the plant floor. This covers tasks such as managing machine capacities and material buffers as well as capturing production data. Coarse-grained planning for these activities is done with the ERP system SAP R/3. Production plans are created in the ERP system and loaded to the MES.

Transportation units play a major role in the information management of the production process because they carry information about the contained materials. CLU uses plastic pallets for outbound shipping. These plastic pallets are piled on larger wooden pallets within stock and for transportation. Additionally, CLU uses metal baskets as internal transportation units. Table 4.2 provides an overview of the employed transportation units.

Two kinds of material flow can be identified on the plant floor from the intake to the assembly. After assembly, the sliding clutches are packed into boxes, which are directly used for shipment. One of these material flows is the flow of ready-bought parts. These parts are directly moved from the intake to the assembly step. At this step, they are loaded to machines for assembly. This is done in the same transportation units in which they were shipped. The other material flow is that of parts that are processed before they are assembled.

Before the first production step, materials that need further processing are unloaded from the plastic pallets on which they were shipped and are loaded

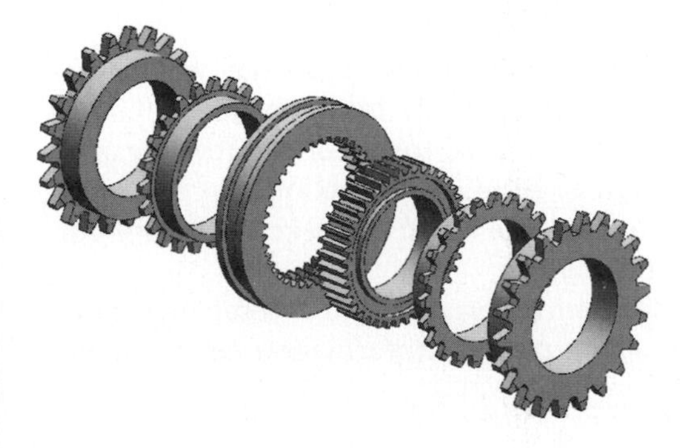

Fig. 4.5 Model of a sliding clutch

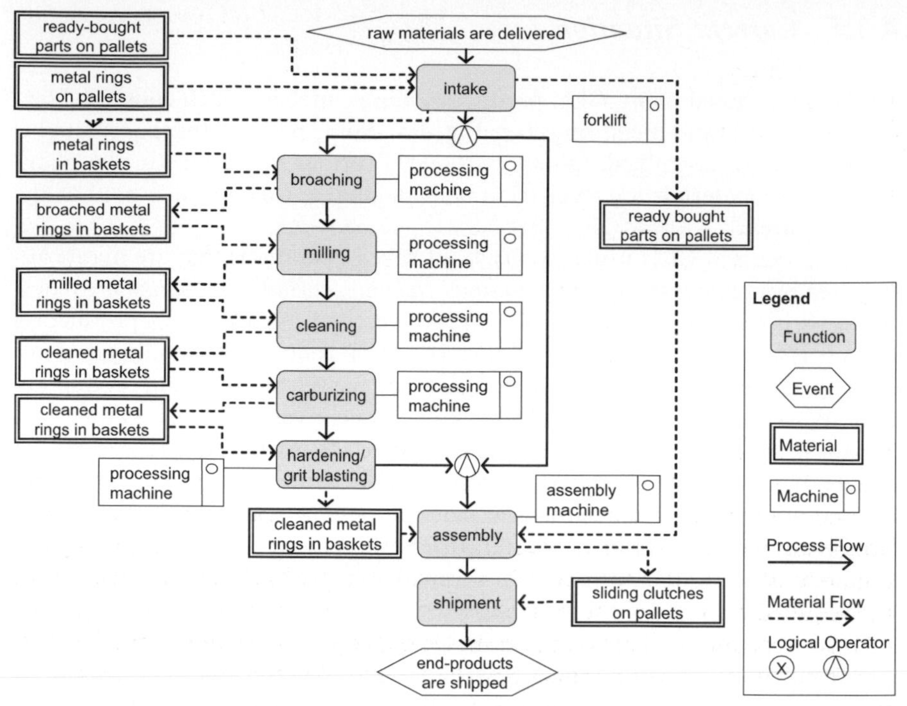

Fig. 4.6 Production process at CLU's investigated plant

Table 4.2 Transportation units at CLU

Type of transportation unit	Use
Plastic pallet	Inbound shipping, loading of assembly machines
Plastic boxes	Outbound shipping
Wooden pallet	Transporting of stacks of plastic pallets
Metal basket	Internal transportation of materials and loading of machines in production

to metal baskets. All machines in the production line are loaded with several stacks of these metal baskets. After processing, the machines place the processed parts back into the baskets that were used for loading. However, parts may be moved from one basket to another during a production step; as a result, the association of one part with a basket or a stack of baskets does not continue throughout the entire production line.

Each stack of baskets is associated with an order via an accompanying document used to document the production process. Information about the ma-

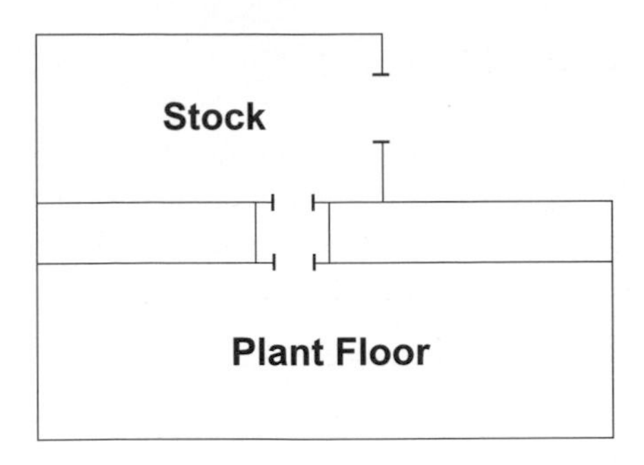

Fig. 4.7 Layout of the plant (not proportionally scaled)

terials in the baskets and the conducted operations are written on these documents. However, material parts may move from one basket to another during an operation. Consequently, information on an accompanying document may not be valid for all parts in the corresponding basket stack. Yet the accompanying documents hold sufficient information to verify that each part has passed the required operations and to identify the type of material in the baskets. All of this information is kept on paper and is not digitalized at any point. Manual, paper-based information management further holds the risk that documents may get mixed up on the plant floor. It is therefore insufficient to use this information for narrowing the scope of recalls.

The accompanying documents are also used to monitor manufacturing progress. Workers are advised to scan a barcode on the documents after the corresponding basket stack has passed an operation. Information about the process status is thereby fed into the back-end system. However, workers sometimes do not scan the documents immediately.

In order to investigate potential RFID application scenarios at CLU, the plant layout must be taken into account (see Fig. 4.7). Locations of scan points depend on how materials are moved across the plant. This influences the total number of required readers and therefore the investment cost for potential RFID solutions.

At the investigated plant, all shipments leave and enter through a single gate. This gate leads to the plant stock. The plant floor is connected with the stock via a second gate. Since all products must pass both gates at the stock, these gates are good candidates for locations of RFID readers.

On the plant floor, machines with identical or similar functionality are spatially grouped. These functional groups are ordered according to the production process flow. Hence, materials move along relatively fixed routes across the plant floor. Readers along these routes could be used to scan passing materials. Because of the functional grouping of machines, waypoints along the route correspond to operations. Thus, a few scan points at the plant floor could be sufficient for monitoring all processes.

4.2.2 *RFID Perspectives*

CLU is currently running tests for a planned RFID implementation. This case study refers to the application that CLU is about to implement, as well as other application scenarios that may be realized in the long term. We identified the following scenarios for the investigated plant, targeting the following four improvements:

1. Improving traceability through bulk scans
2. Automating warehouse management
3. Improving production tracking
4. Improving customer service

4.2.2.1 RFID Application Scenario #1: Improving Traceability Through Bulk Scans

The predominant reason for considering RFID solutions at CLU is to improve the traceability of the production process. The goal is to narrow the scope of possible recalls and to reduce resulting costs. If a production error is detected, all clutches that are potentially affected must be checked manually. If those faulty products were already shipped, checks must even take place at the customers' plants. This results in additional costs for sending workers to the customers. These costs for recalls are even higher for clutches that were already used in a customer's production. In this case, CLU must pay penalties. These penalties amount to about 1.5% of CLU's total business volume.

Because of the high cost for supplementary checks and penalties, reducing recalls holds the potential for significant savings. Costs for recalls could be reduced if tracking information would allow narrowing of the scope for checking. Currently, one application scenario for recoding tracking information with RFID is being evaluated in cooperation with a consulting firm. Using this tracking information would help limit the scope of recalls and checks. In this scenario, all plastic pallets carrying material parts will be equipped with RFID tags. The idea is that suppliers will provide information about the shipment by using these rewritable tags.

Two scan points are planned to monitor three events within the production process. The first event is captured at the gate between the intake and the plant floor. Recording this event allows CLU to track which items are currently in the production cycle. Readers at the assembly station register the second event. Each plastic pallet is scanned within the used assembly machine. (Note that only ready-bought parts are transported on pallets at this stage.) Thereby, for each pallet of clutches it is recorded from which pallets the material parts were taken. The last event is captured when the transportation units pass through the gate between the intake and the plant floor again. The semantic for this event is that the production cycle has been finished.

Barcode labels could theoretically be used in the same way as RFID tags without user memory. However, applying barcodes is practically infeasible because CLU wishes to scan the pallets in bulk at the gate of the plant floor. Through this gate, whole stacks of pallets are transported on a forklift. All pallets on a stack could be scanned in seconds if RFID were applied, whereas barcode solution would require time-consuming manual scanning, rendering a barcode approach inapplicable.

Automatic bulk scanning demands RFID tags with a long range. The plan is to scan bulks of plastic pallets when they are moved between the stock and the plant floor. Here, the plastic pallets are transported on wooden pallets. A read range of at least 2 meters is desirable due to the size of the wooden pallets. UHF tags corresponding to the Gen 2 or ISO 18000-6C standard fulfill these requirements. These two standards are compatible and dominant for application in logistics. Another argument for these tags is the relative low price of less than 30 cents each.

In the long term, CLU should also consider tagging the metal baskets. These baskets are used to load and unload the machines as well as to transport parts to a subsequent operation. Tagging these baskets would enable more accuracy in tracking the production process. With RFID readers at each machine, the baskets could be traced trough the whole production process. As result, recalls could be narrowed down further. However, this application scenario would require an update of the machine/PLC software. With the current software, some parts get moved from one basket to another. Consequently, tracking on the basket level is not feasible with the current software.

Using RFID tags that can cope with the influence of metal must be considered for possible tagging of the metal baskets. Possible obstacles are shielding effects or detuning of the radio signal. Here, only a field study could fully clarify whether the metallic surroundings would allow use of RFID.

4.2.2.2 RFID Application Scenario #2: Automating Warehouse Management

CLU had already conducted initial tests to see how RFID could improve traceability of the production process. However, realizing RFID application scenario #1 would leverage other scenarios in which RFID could deliver benefits. Tracking the production process requires the transportation units to be equipped with RFID tags. This includes the transportation units that are also used for shipment processes. RFID tags on these transportation units could be reused at the warehouse. For example, it would be possible to use the RFID tags for managing the warehouse's inventory. Either smart shelves (shelves that are equipped with RFID readers) or mobile readers could be used for quick updates of the whole inventory. Compared with barcode-based solutions, this would reduce the workload for warehouse management and could improve the accuracy of the inventory list.

Furthermore, goods entering and leaving the warehouse could be automatically registered if RFID readers were installed at the warehouse entry. For instance, shipping notices could automatically be compared with received goods. Regarding the RFID hardware, the same setups as in common logistic applications can be used (for instance, by using UHF RFID tags of the Gen 2 standard). Typical gate installations with about four antennas could be used for readers.

4.2.2.3 RFID Application Scenario #3: Improving Production Tracking

Currently, CLU uses barcodes to measure the progress of running production tasks. The barcodes are printed on documents that accompany the stacks of metal baskets that carry the materials for processing. Workers are advised to scan the barcodes after the corresponding stack has passed a production step. Information about the progress of the production is thereby fed into the back-end system, and the progress is documented. Yet workers do not always scan the barcodes on time. Sometimes the accompanying documents are not even collected and scanned until the end of a shift. Thus, the status information obtained by scanning barcodes does not always reflect reality and is of limited use for performance analyses.

Measuring the production progress could be automated if scanning were done with RFID instead of barcodes. This would require equipping every stack of internal transportation units with an RFID tag. Then readers at each production step could detect whether a stack has been moved to the next station, and the information could automatically be updated in the back-end system.

Furthermore, the production process could be monitored in greater detail by implementing additional scan points at the plant floor. For instance, for each stack of metal baskets, the amount of time spent at each process could be recorded. This would allow CLU to further narrow recalls and monitor the process in greater detail. But if scanning is not automated, the employees' workload would increase. Compared with barcodes, RFID allows for automated scanning in more situations and thus more data collection. Applying RFID on the plant floor could help CLU collect more detailed information about its production process in real time.

This additional information allows for better analysis of the plant's efficiency. For example, lead times and material buffers could be measured in detail between each scan point. Analyzing this information may help determine potentials for improvements. Also, exceptions in the process flow could be detected. In this context, CLU may also consider collecting additional sensor data on the plant floor. Evaluation of this data could help detect unusual and undesirable situations at each operation. Performing these analyses on real-time data allows for fast reaction to exceptions. Potentially negative outcomes of unexpected events could thus be limited or even completely avoided.

How many scan points on the plant floor would be desirable depends on the investment cost and the expected gains in efficiency. The cost for each scan point must be traded against the anticipated benefits of the collected data. A recall of a single clutch costs about 5,000 euros. This is relatively high compared with the cost of RFID readers (currently about 300 euros for mobile readers and about 3,000 euros for stationary gates) and autonomous sensor tags (currently about 100 euros). The low-cost tags of the Gen 2 standard (which were recommended above) could also be used for automatic registration of objects at checkpoints. However, the cost for adapting the IT system must be considered as well. A positive return on investment (ROI) is likely if additional data could help reduce recalls significantly. More detailed information about the causes of production errors is needed for a well-founded recommendation in this context.

4.2.2.4 RFID Application Scenario #4: Improving Customer Service

The application of RFID at the production plant may not only improve production itself but could also enable better cooperation with customers (typically OEMs). In this context we distinguish two ways of cooperation.

First, RFID could be used to improve information exchange with customers. Here, RFID may serve as both the medium for information exchange and the enabler for gathering the information provided to the customer. As discussed above, applying RFID in the production process would enable the recording of fine-grained process information in real time. This information may be used by CLU but could also be shared with the customer that ordered the products currently in production.

Knowing about the status of products at CLU would likely make it easier for the customer to plan its own activities. For example, uncertainty about lead times could be reduced, which could improve scheduling of the customer's production. Furthermore, the availability of fine-grained real-time information enables timely reaction to exceptions and fast adoption to changes in performance. However, such real-time information could be shared only via a network connection with the customer. Thus, the RFID infrastructure on the plant floor would need to be integrated with this network for data exchange, and before the data are passed to the network, it must be filtered and aggregated.

Second, another way to improve customer service is to enable the shared use of RFID tags. The tags that CLU is using on transportation units (pallets) in the outbound could also be used for automatic scanning at the customers' inbound. Using the tags would be an option for exchanging information that is tightly coupled to a certain object. An example of such information is historic data about the production of the respective object. Thus, CLU would enable possible improvements at the customer's side by equipping the transportation units with RFID tags. In such a scenario, CLU might also discuss cost-sharing models for RFID tags with its clients.

Using RFID for data exchange requires that the applied tags have sufficient rewritable memory and that the cooperating partners can agree on a tag standard. Currently, the Gen 2 standard is the dominant one for UHF tags.

It is compatible with the recently announced standard ISO 18000-6C and is supported by early RFID adopters such as Wal-Mart and Metro. Therefore, we recommend using Gen 2 tags because compatibility with other UHF RFID systems is most likely if this standard is used. Also, the Gen 2 specification allows for integrating rewritable user memory, which could be used to exchange information with the customer.

Exchanging data via RFID tags has the advantage that no network infrastructure is needed. This reduces the costs for establishing and maintaining the IT infrastructure required for network-based information exchange. Another advantage is that the information is shipped together with the object it belongs to. This could reduce lookups in databases or via a network and allow some information processing to be performed locally without interaction with the back-end system. For instance, objects may be directed to certain process steps based on the information on the respective tag.

4.2.3 Costs and Benefits

In the last section we described four potential RFID benefits at CLU. However, for some of them it is not possible to calculate monetary benefits without further analysis. For instance, gains due to improved production tracking and improved customer service cannot easily be quantified without further analysis.

The main driver for implementing RFID at CLU is the expected savings from improving the traceability of products. If a production error is detected, all clutches potentially affected must now be checked manually. For products already shipped, checks must even take place at the customers' plants. This results in additional costs for sending engineers to the customers. Costs for recalls are even higher for clutches that were already used in a customer's production process. In this case, CLU must pay penalties, which now add up to about 7.5% of CLU's revenue. Thus, improving traceability and thereby narrowing recalls could lead to significant savings.

The cost for realizing this potential depends on the desired level of granularity. The basic application can be implemented with only one reader gate at the plant floor and reusable tags on the transport units. Yet the tracking would still be relatively coarse-grained. To achieve tracking on the level of individual operations, additional readers on the plant floor would be needed. Furthermore, the software of the machines would need to be updated as the machines move clutches within and between transportation units.

If RFID tags and readers are in place, this infrastructure can be used to improve warehouse management and speed up the processes for loading and unloading shipments. Here, labor costs can be saved, and the overall process can be accelerated. In combination with the scenario for improved traceability, only minor additional hardware costs would be incurred for this application.

4.2.4 *Summary*

In this case study we discussed potential benefits that CLU may gain from RFID adoption. To specify these possible advantages, the current situation and production processes were investigated. We identified areas that could be improved by using RFID and outlined corresponding application scenarios. This comprises an investigation into whether the use of RFID in this production process could be advantageous compared with the current application of barcodes.

In addition to the outlined RFID potentials in the plant itself, business processes across the supply chain could be improved further to fully exploit the advantages of an RFID-enabled supply chain. Possible measures include selective communication of RFID data to CLU's customers and business partners. At this point, the production data are simply written into a back-end system. The strategic issues to be chiefly considered are quality and customer relations. OEMs will focus increasingly on traceability, which will provide a window of opportunity for suppliers that are well prepared for such demands. In the longer term, a tighter IT-enabled integration of the entire supply chain seems likely. At this point, however, the implementation of such integration measures, including related security measures and standards activities, is a costly and possibly risky endeavor. Possible benefits may not be distributed fairly between the different parties in the supply chain. This concerns not only costs for RFID readers, tags, and required software but also costs for changing the underlying processes.

4.3 COO: Engine-Cooling Modules

COO produces engine-cooling modules and air conditioning devices (see Fig. 4.8). It is a first-tier automotive supplier. Currently, COO applies RFID in two assembly lines for tracking cooling modules along the assembly process. RFID was introduced because of a customer's demand for better documentation of the production process. While the current setup serves to gain some

Fig. 4.8 Model of an engine-cooling module

general experience with the technology, so far there have been no detailed ROI calculations.

For COO's customers—all OEMs—long-term traceability of their products is very important. This includes parts and components that COO purchases from second-tier suppliers. However, using RFID implies investments and modifications of existing production and business processes. At this point, it is not clear whether the resulting improvements justify those investments.

This case study investigates (1) whether RFID is currently used in an efficient manner and (2) whether COO can expect competitive advantages from introducing RFID on a broader scale.

4.3.1 Current Situation

In 2003 COO built two identical RFID-enabled assembly lines that produce engine-cooling modules. Both lines work in parallel to fulfill one order. They are guarded by a JIS control system. The JIS software is responsible for the correct sequencing of the engine-cooling modules into the pallets, which are later loaded onto the truck to the customer.

The two assembly lines produce 1,300 engine-cooling modules per day, operating 12 hours per day, 5.5 days per week. The engine-cooling modules are put onto pallets in the JIS order specified by the customer. After production they are instantly loaded into a truck and shipped to the customer. One order can require varying amounts of different types of coolers. In total, up to 32 different engine-cooling modules can be produced.

The engine-cooling modules are mounted on carriers (one module per carrier) and moved along the assembly line from workstation to workstation. Each line is configured dynamically to be composed of up to six serially connected workstations. At every station, workers perform a certain number of tasks. Tasks are not assigned to workstations in a fixed manner, which allows for some flexibility in planning the process. It is possible to get by with three workstations if necessary (for example, if workers fall ill). In such cases, the number of tasks performed at each workstation increases, and throughput decreases. COO finds the optimal number of active workstations to be five per line.

The complete process is coordinated by two PLCs (manufacturing control software) and the overlying JIS software. The JIS software controls the correct sequencing of the engine-cooling modules into pallets, which are later loaded onto a truck. One pallet carries six engine-cooling modules. When one line assembles all modules for the current pallet, its PLC sends a request to the JIS. Subsequently, the JIS advises the PLC about what kind of modules it has to fill into the next pallet. One PLC coordinates one production line, where it is responsible for the assembly. The PLC produces a data set, including type information and job parameters for each engine-cooling module to be assembled. The data set is specific for each workstation involved. This data set is stored on an RFID tag that is attached to the carrier holding the module. As the engine-

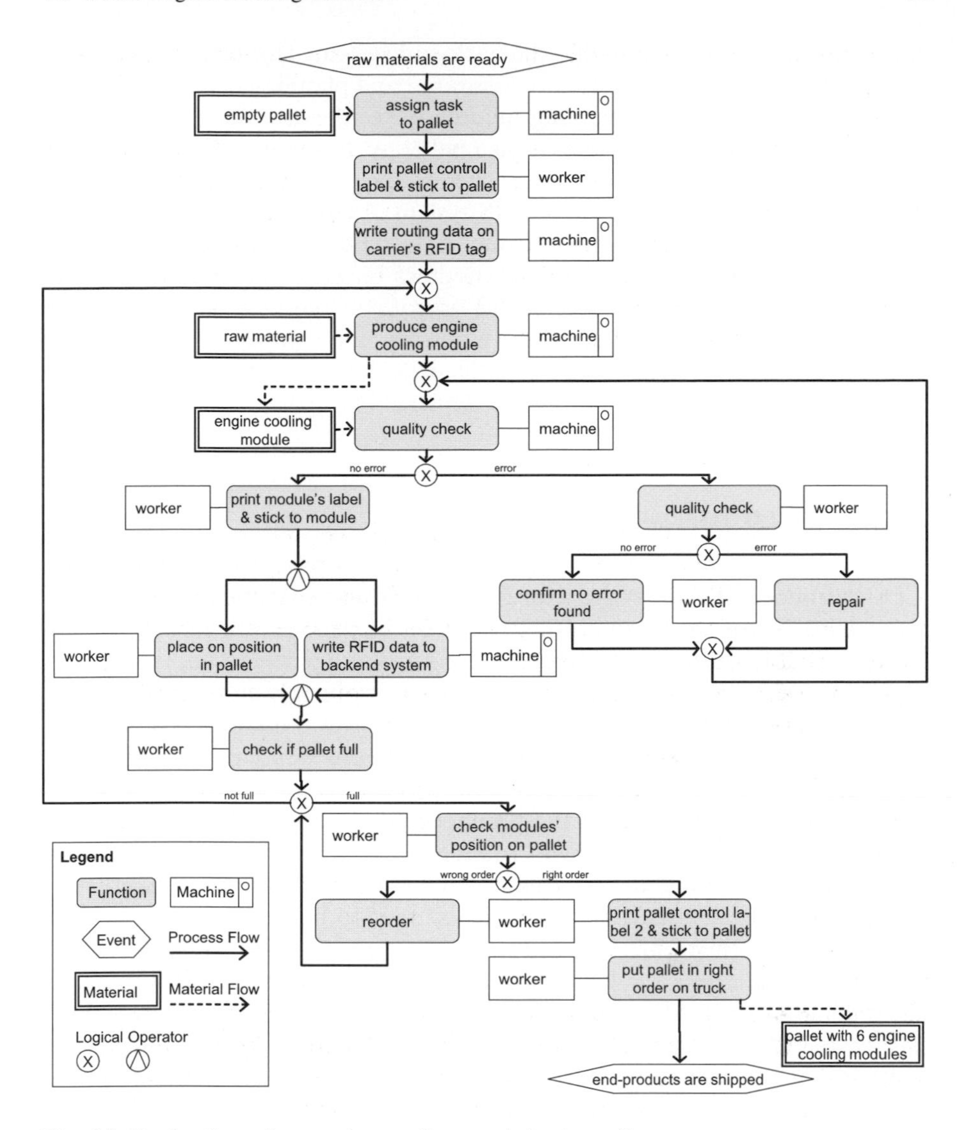

Fig. 4.9 Production of an engine-cooling module at one line

cooling module moves along the assembly line, the information on the tag is constantly updated.

At each workstation, job instructions are read from the RFID tag and are used for local assembling purposes. Once the worker has completed the job, the relevant information is written back to the tag. For each engine-cooling module, about 3,000 bytes of data are saved on tag. These data include which cooling module was produced, how the production steps were distributed among the active workstations, and whether all tasks were performed accurately. Errors made by the robots are also recorded. At the end of the production cycle, the complete data set is stored in the back-end system. This recorded information is made available to the quality assurance and planning department.

When a finished engine-cooling module leaves the assembly line, a worker lifts it from the carrier and places it on a pallet. At this point all the data on the RFID tag (which—remember—is attached to the carrier, not to the module itself) are transferred to the back-end JIS system. A barcode label with the module's serial number is printed and affixed to the module. When the pallet is full, the worker checks the position of all modules on the pallet. When the check succeeds, a control label for the forklift operator is printed and affixed to the pallet. The forklift operator scans the barcodes on the paper label, determines with this the pallet's position, and puts the pallet into the correct position in the truck. The whole production process is depicted in Fig. 4.9.

4.3.2 RFID Perspectives

COO introduced RFID at the request of one customer that expects improved traceability of the produced parts. Traceability of parts is of increasing importance to OEMs, partly due to legal requirements and partly to improve quality in the long term by identifying faulty components quickly and reliably. Once a part is separated from the carrier, the information is transferred to the back-end system. The carrier and the RFID tag attached to it can be recycled immediately.

Separating the product and the product-related data implies to some challenges regarding data management. In order to obtain accurate data on a given engine-cooling module, the module's ID must be available and forwarded to COO's relevant databases. These databases need to be maintained to provide the relevant information in the long term. In contrast, having the required data stored with the respective object ensures easy and consistent access throughout the lifetime of the tag (which, ideally, is identical to the lifetime of the object). Moreover, if the tags were attached to the engine-cooling modules directly and remained on the parts after delivery, the OEM could read the information upon receipt. Barcode scans would not be required, and bulk reads would be possible. This could improve overall productivity of the supply chain.

At the moment, no production data are passed on to the ERP system because the information is considered to be of operational value only. Production and business processes across the supply chain could possibly be improved further to fully exploit the advantages of an RFID-enabled supply chain. This would require tighter integration of the RFID data into the existing ERP infrastructure. A closer integration into the ERP system could also facilitate the creation of a data warehouse to document the production process over a longer period of time. Customers may be interested in obtaining these data. This may even account for new business opportunities. For example, Metro AG, Germany's largest retailer, already realizes substantial revenues by offering sales data to its suppliers.

4.3.3 Costs and Benefits

Because the RFID tags are reused, costs for the described application scenarios are fixed. Thus, the number of RFID tags remains small, and tags with extended storage are affordable despite their relatively high costs. In a barcode-based solution, more functionality in the back-end system would be needed for realizing the same application. By using the RFID-based solution, investments in the back-end system could be reduced. Also, COO's production control system is less complex because the data can be written to the tag and updated "in bulk" at the end of the process. Perhaps most importantly, COO has implemented the RFID application because of the demand of an important customer.

4.3.4 Summary

This case study presents an analysis of the productivity and potential of RFID at COO, a first-tier automotive supplier. The key question was whether RFID is currently being used in an efficient manner.

RFID was installed on the demand of one customer, which seems to be satisfied with the results. The RFID technology allows recording of detailed information about the assembly of each single engine-cooling module, thus providing thorough documentation of the production process and thereby a reliable basis for tracing parts in the case of future breakdowns or recalls. However, the advantages of using RFID versus a classic barcode solution are limited. We want to point out here that business processes across the supply chain could be further improved in order to fully exploit the advantages of an RFID-enabled supply chain. Possible measures include the use of part-specific RFID tags that remain on the part after delivery, tighter integration of the RFID data into the existing ERP infrastructure, the selective communication of these data to COO's customers and business partners, and the implementation of a production data warehouse.

4.4 CAS: Production of Cast Parts

CAS, a manufacturer of cast parts, produces steel cast parts of various sizes. The models for creating the cast molds are designed and produced at CAS, and the availability of the models is crucial for operations at the plant. To ensure fast retrieval of models from stock, CAS plans to implement a tracking application that keeps accurate information about the position of each part of the model. CAS expects that RFID could be used to automatically update model positions despite a problematic (dirt, metal) production environment.

4.4.1 Current Situation

The products of CAS are casted parts in diverse sizes (see Fig. 4.10). The raw materials used are discarded metal parts that are melted and casted into the desired shape. CAS also designs the models required for creating mold forms for the casting.

The overall production process can be subdivided into nine separate steps (see Fig. 4.11). The production process starts with scheduling of the production tasks. Currently, the casting process is planned on a weekly basis. Thus, all needed assets and materials are known several days in advance. After a schedule for operations is determined, the corresponding model parts for the production task are fetched from stock. Models are the central asset in the production of cast parts. They consist of two model covers and up to 20 core forms. The covers and the cores are stored independently in one of the plant's numerous storage rooms. Each part must be fetched in a timely manner before production can start.

After the required model cores and model covers are fetched, they are used to create sand forms. That is, the model parts are mounted to a special casing, and a mix of sand and glue is injected into them. The resulting sand forms are negatives of the final product. Subsequently, the sand forms are used to cast metal into the desired shape. When the cast part is cooled, the sand form collapses. Finally, excess metal at parting lines is removed, and the product is shipped to the customer.

Because model parts are the core asset of the production process, efficient management of these parts is important. CAS keeps track of about 5,000 models. These models consist of about 10,000 model covers and about 20,000 model cores, which are stored separately. About 80% of the models are actively used in production and are therefore repeatedly moved between various locations in the plant.

Currently, positions and movements of the models parts are managed via the SAP R/3 ERP system. At the beginning of production, the workers select all needed model parts from stock. If a part location is unknown, workers must search all possible locations. At the investigated plant, searching all locations where a model part might be is a very time-consuming task. The plant is lo-

Fig. 4.10 Model of a cast part

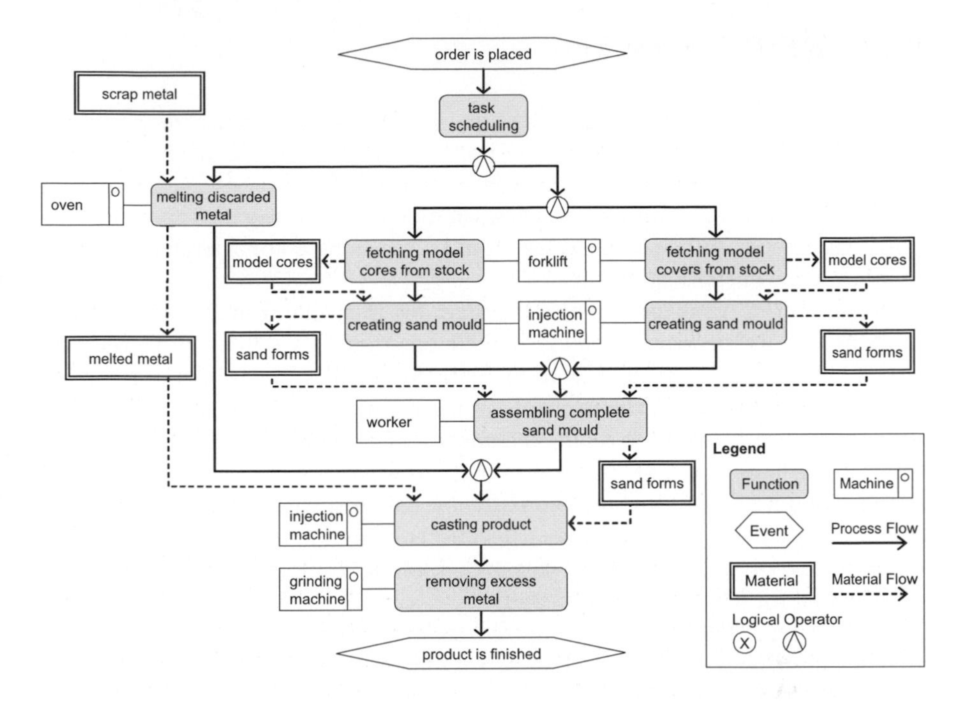

Fig. 4.11 Production process at CAS

cated in a building that includes older building parts that were not specially designed for their current use. Because of this, it is not possible to use a central stock. Instead, several storage rooms are distributed across the building. Altogether, model parts can be at one of the stocks in the plant, at one of the repair facilities, or at a material buffer on the plant floor.

For keeping track of the model parts, the workers are requested to update the positions of these assets in the SAP R/3 system. However, updating the positions is a cumbersome manual task and is therefore not done consistently. According to CAS's workers, a consequence is that about 2% of all model parts need to be searched for before they can be used in production. Reducing search times is of special interest to the company because searching is a time-consuming task that is likely to delay production.

CAS addresses this problem by planning time buffers to allow fetching of the required model parts. Currently, production plans are created on a weekly basis, and model parts are fetched the day before they are needed. This allows time to react when parts must be searched for. However, because CAS is about to switch to planning on a daily basis, leaving less time for searching, the CAS IT managers expressed the explicit desire to improve the tracking system for model parts. This would reduce search times and make short-term planning more reliable.

4.4.2 RFID Perspectives

CAS is currently investigating how to realize a tracking system that would enable reduced search times for model parts. In particular, the accuracy of the current solution needs to be improved so that the location of each part is at least roughly known. Employees at CAS stated that knowing the room in which a lost part needs to be searched for would already be a considerable improvement over the current situation.

At present, CAS is using its ERP system for planning movements of model parts as well as for keeping track of their positions. By default, it is assumed that all movements are conducted as planned. That is, the model parts' positions are updated according to the work plans for transportation. If the actual movement differs from the planned transport, workers can update the data records in the back-end system via terminals. However, workers neglect necessary manual updates, resulting in inconsistency between the stored position data about model parts and their actual locations in the plant. As mentioned, the consequence of such inconsistencies is that about 2% of the model parts must be searched for when they are needed.

CAS expects RFID to help improve its management of model parts such that search times could be avoided or significantly reduced. The company expects RFID to be used for an application that automatically updates the positions of model parts in the back-end system. This would avoid the problem of workers neglecting to update the database and would ensure consistent information about the positions of all model parts.

Several options need to be considered for the technical realization of such an application. We subsequently describe two scenarios for the investigated plant. They target the following possibilities regarding ways to reduce the search time for model parts by using RFID:

1. Tracking model parts using fixed reader gates
2. Tracking model parts using mobile readers

The two scenarios differ not only in the hardware used but also in the requirements for the software and the achieved granularity of tracking information. We discuss both options in more detail.

4.4.2.1 Tracking Model Parts Using Fixed Reader Gates

One option to track model parts with RFID is by using fixed reader gates. In this scenario, reader gates would be installed at key waypoints on the plant floor. That is, it must be ensured that at least one gate is passed when a model is transported between different locations. Workers at the investigated plant estimate that about 10 reader gates at certain room doors would be sufficient for ensuring that at least one gate is passed during each transportation task. Note that model parts could be tracked only at room granularity this way. Yet this would be a significant improvement compared to the current situation. If a model part has been misplaced, the search time could be reduced if workers know which room the item is in.

For this option, each model part would need to be tagged to one or more RFID tags. Each part could be labeled with multiple tags aligned in different angles and situated on different sides of the model part. This redundancy would generally increase reading reliability. However, a large number of tags may cause collisions on the communication channel. If the number of collisions is too high, the model part may need to pass a reader gate slowly in order for all tags to be captured. However, only a few model parts would be transported at the same time at CAS. Thus, problems due to collisions are not likely to occur, even if workers pass the gates at normal walking speed.

Concerning the frequency, UHF is likely to be the most suitable radio frequency spectrum for the desired application. This is because UHF typically enables higher ranges compared with HF- or LF-based RFID solutions. Reader gates at CAS would need a diameter of about 2–3 m to conveniently fit the model parts through the gates. Therefore, read ranges of several meters are desirable.

4.4.2.2 Tracking Model Parts Using Mobile Readers

Another option for implementing a tracking application at CAS is to use mobile readers. Here, each worker in charge of transporting model parts would be equipped with a mobile reader. For this solution, fewer readers would be needed at CAS compared to the scenario above, in which stationary readers are used.

In a tracking application with mobile readers, moved model parts would be registered at the location where they are dropped. RFID can help automatically capture both the location and the identity of the respective model part as it is dropped. As in the scenario with fixed reader gates, the model parts would

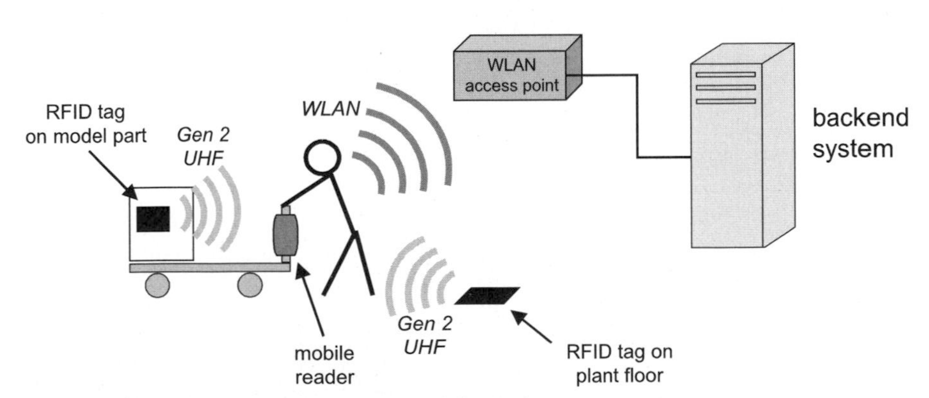

Fig. 4.12 Tracking model parts using mobile readers

be labeled with one or more tags holding an identifier. Thereby, moved model parts could be easily registered with the mobile readers. The location could be determined using RFID tags for positions. Position tags could be distributed in the plant, such as on the plant floor (see Fig. 4.12). Note that the granularity of this tracking solution depends mainly on the number of applied position tags. Because low-cost tags can be used to mark positions, it is financially feasible to achieve high granularity. Thus, at CAS the application scenario with mobile readers would facilitate more fine-grained tracks than the scenario with fixed reader gates.

When a model part is dropped, the model ID and the position tags nearby would be read out by the mobile readers. The back-end database could thus automatically be updated with the model part's position.

For determining whether a model part is at a new location, two options can be considered: First, workers could manually issue RFID readouts when they drop a model part. This poses the risk that workers would occasionally forget this task. Second, the company could use advanced software solutions that automatically infer whether a model part has been dropped. This inference could be made using a record of read events that are periodically captured. Smart read devices may run software that performs the described inference locally. In this case, tracking data that have already been evaluated can be communicated to the back-end system, and single read events could be dropped. Otherwise, if the evaluation is conducted in the back-end system, all captured read events must be sent to the back end.

Communicating the record of read events to the back-end system may also be done in two different ways: If delays of up to several hours are acceptable, the buffers in the mobile readers can be loaded to the back-end system via PC-based cradles. Workers could put their mobile readers into the cradles on the plant floor from time to time. Alternatively, live updates could be realized using WLAN technology. Some mobile readers can optionally be equipped with WLAN connectivity and can thereby be permanently linked to the back-end system. This would allow the locations of model parts to be updated in real time.

4.4.3 Costs and Benefits

CAS is considering the application of RFID to overcome its problems in tracking model parts. If these parts cannot be found in time, the production process stops. Fast retrieval of model parts will especially be a challenge when CAS realizes its goal to switch from a weekly planning cycle to a daily planning cycle. Therefore, CAS must evaluate the cost of an improved tracking system versus putting more effort into the search process.

To determine the hardware cost, two different scenarios must be considered at CAS. One scenario is designed for tracking model parts on the room level. For this scenario, about 10 reader gates would be required at waypoints on the transportation routes. Additionally, several tens of thousands of RFID tags would be needed to label the model parts. In the other scenario, mobile readers and position tags on the plant floor are required. Excluding spare readers, two mobile readers would be sufficient to equip each of the two warehouse workers. Compared with the first scenario, the second scenario needs fewer investments in hardware, although additional RFID tags for marking positions on the plant floor are needed. Besides being cheaper in terms of hardware costs, the second scenario would allow tracking with finer granularity. However, evaluating the reader data is more complex in the mobile reader scenario. Thus, more complex and, probably, more expensive software would be required.

4.4.4 Summary

In this case study we investigated potential solutions for tracking model parts at CAS. We described the benefits CAS may derive from implementing a tracking application and the technological options that may be considered for the implementation. We elaborated on different hardware solutions for capturing the tracking data. In particular, we discussed potential implementations with mobile readers and fixed reader gates. For these scenarios we pointed out differences such as the achievable granularity and the related hardware costs. We furthermore described how data can be captured, evaluated, and communicated to the back-end system.

Before one of the options is implemented, hardware tests need to be conducted to determine the read reliability that can be achieved in the different scenarios. Results of these tests have implications not only on the hardware setup but also on the needed software functionality. The achieved reliability and the desired degree of automation determine what inference and filter operations need to be performed on the data. Finally, after these issues are determined, the appropriate architecture and distribution of functionalities can be chosen for the tracking application at CAS.

4.5 CON: Production of Electronic Connectors

CON processes metal and plastics in order to produce electronic connectors. CON was interested in a general study of whether and how RFID could improve operations on its plant floor. This was partly motivated because CON faces increasing competition with manufacturers in low-wage countries and needs to increase its productivity. In this case study, we investigated several scenarios of how RFID can be applied in CON's production processes. In some of these scenarios, barcode technology could be an attractive alternative to RFID-based solutions. We will discuss the relevant trade-offs.

4.5.1 Current Situation

In this section we discuss the current situation at CON's plant, based on a field visit to one specific production plant in April 2006. This includes a detailed description of the production processes for connectors as well as an investigation about the internal material and information flow. This section also deals with the handling of containers during the production process. Here the focus lies on the process of packing and labeling.

4.5.1.1 Production

The specific plant produces different types of electronic connectors (Fig. 4.13). Figure 4.14 depicts a generic model of the production process, showing how raw materials and products are processed in different production steps. The production can be divided into six steps: *punching*, *galvanizing*, *molding*, *injection molding*, *assembly*, and *shipment*. Each step may consist of several substeps. We categorize the production steps into six groups, which is sufficient for our case study.

The major raw materials for production are copper straps and plastic granulates. Internal transfer of materials is done using coils for copper straps and boxes for molded plastic parts. The processed copper straps follow one of two different routes in the production; the route depends on the types of end products for which these straps are intended. Either way, the first production step is *punching*. Here, the respective machine punches the desired pin shapes into the straps. After the copper straps are punched, they are reeled on empty coils and sent to the second production step, *galvanizing*. Galvanizing is performed to protect the copper against corrosion.

After galvanizing, the processed copper tapes are sent either to the production step *assembly* or to *injection molding,* depending on the connector type that will be produced. If the processed copper straps are sent to *assembly*, a machine automatically assembles the straps with molded plastic parts. Molded plastic parts are made from synthetic granules in the *molding* step. The molding step serves to mix and shape various raw synthetic granules depending on diverse product specifications. The output of the molding step is always sent to

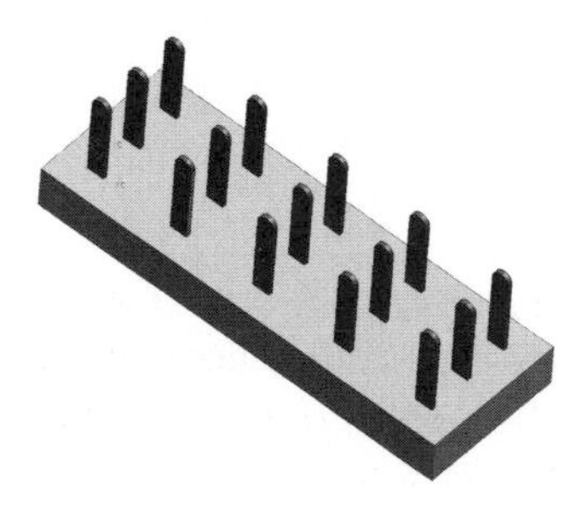

Fig. 4.13 Model of a connector

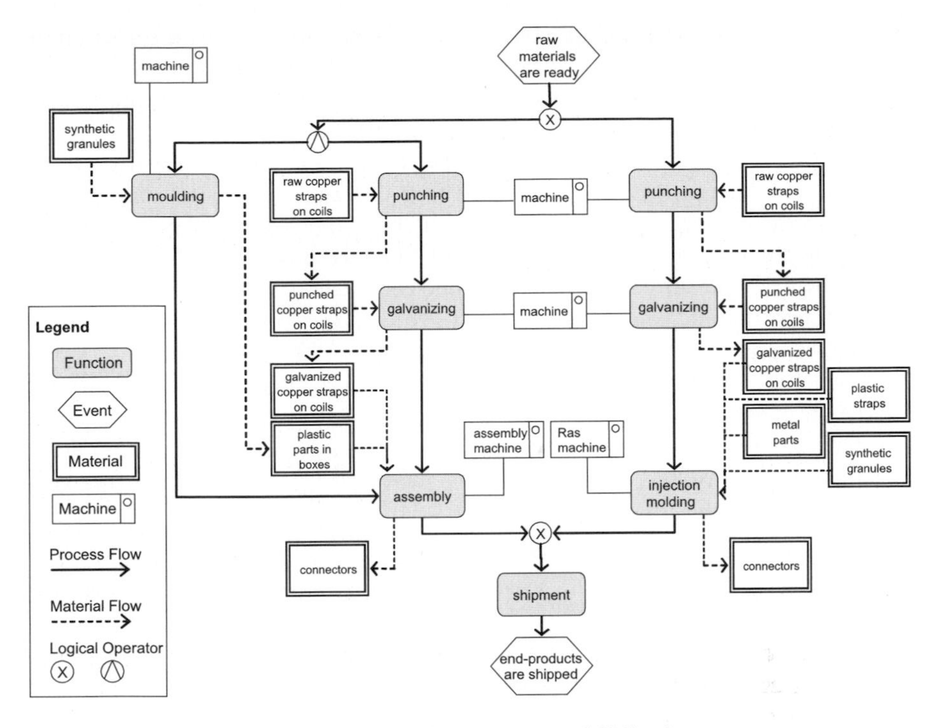

Fig. 4.14 Generic model of the production process at CON's plant

assembly. After the assembly step is completed, the end products are sent for *shipment*.

If the galvanized copper straps are sent to *injection molding*, they are processed by a machine called the Ras machine. Besides the copper straps, plastic straps on coils, metal parts, and synthetic granules are other raw materials used in this step. The Ras machine mixes and shapes the synthetic granules and assembles these shaped parts with the other two raw materials and the galvanized copper straps. The output of the Ras machine is a finished product and is directly sent to shipment without further processing.

Regardless of the production step from which the finished products arrive, they are brought to shipment. After each production step, it is possible to forward the processed materials to storage, which is located in the middle of the plant's building.

4.5.1.2 Internal Flow of Materials and Information

Internal transfer of materials is done with material-specific transportation units: coils for copper straps and boxes for molded plastic parts. Thus, there are two types of internal transportation units in the plant. Regardless of which one is used, accompanying tickets are employed to record information about the carried materials in the transportation units. These tickets include information such as the type and number of contained materials, the production date, and the charge that the the materials belong to. Using information on the tickets, the required materials are selected for the next production step.

In each production step in which copper is processed, the straps are wound off a source coil, processed, and reeled to a destination coil. Before they are moved to the subsequent step, information about the copper strap is manually written to the accompanying ticket and attached to the destination coil. Figure 4.15 shows an accompanying ticket attached to a coil.

However, all materials and the status of the production for each charge must be tracked in the back-end system of the plant's IT infrastructure. For this purpose, workers are instructed to update the data in the back-end system with the information about the current status of all processed materials and how far the product has been assembled. This is supposed to be done after each completed production step. Workers use terminals on the plant floor for interacting with the back-end system.

4.5.1.3 Challenges in the Internal Flow of Materials

Ideally, the transfer of processed parts to the subsequent step starts with the booking of the current step into the back-end system. Information management is the same for each production step in the plant except for injection molding and shipment. The company assigns great importance to exact tracking of materials and production in the plant. However, there is no mechanism to guarantee synchronization between booking and the transfer of materials to the next step.

Incomplete bookings cause workers in subsequent production steps to interrupt their tasks to belatedly complete the missing booking of a previous step. The time loss per missed booking can be up to 30 minutes, according to interviews with CON's staff. However, instead of correcting a missed booking, it is also possible to proceed with the production. Theoretically, the whole production can be executed without booking until all the production steps are

Fig. 4.15 Coil with ticket

completed. In such a case, the production and the material flow are not tracked accurately.

Another potential for improvement concerns the maintenance of process data. Based on employees' estimations made during our field survey in CON's plant, workers allocate approximately 15% of their working time to maintaining process data. One-third of this time accounts for maintaining tickets for the internal transportation. As mentioned above, the maintenance of tickets includes manually writing information about the processed materials to the accompanying ticket. Automation of this task would potentially increase worker productivity in the plant.

Furthermore, the manual maintenance of the tickets causes errors in data maintenance. According to CON's employees, the error rate is approximately 3–4%; that is, in 3–4% of all situations, a worker writes the required information to the ticket incorrectly. Such errors affect the subsequent production steps because the incorrect information misleads the workers in those steps; the workers search for the required materials and execute the subsequent production steps incorrectly. Moreover, workers in subsequent steps sometimes read the manually written information incorrectly, even if no write errors have occurred. This potentially causes the same consequences as a write error.

If workers write or read the information on tickets about materials incorrectly, machines are possibly configured wrong. The resulting incorrect processing leads to waste and productivity losses.

Another challenge regarding the internal material flow concerns the correct matching of coils and copper straps. As mentioned before, copper straps are wound off a source coil at each operation, and the processed parts of the straps are reeled to a destination coil. However, in some cases the destination coils do not match the processed copper straps in size. If such a problem is encountered, production stops until workers fetch an empty coil with the proper dimensions. This results in a slowdown in production and a decline in the respective worker's productivity.

4.5.1.4 Container Management

At CON, containers are used for shipping products to customers. Here, two groups of containers can be distinguished: reusable and nonreusable. If a container is reusable, it may belong either to a customer or to CON. Some of CON's containers are only temporarily rented. The containers vary according to requirements of the customers and the produced connectors. Reusable containers may be in four different locations: at the packing station, at cleaning, in stock, or at the customer. Until now, CON has had no practical way of finding containers easily.

At CON, almost all of the end products are carried in small packing units, which are packed into bigger containers afterward. All packing units and containers have barcode labels. Furthermore, paper labels are affixed both to the packing units and to the containers. Because every customer may have different requirements for barcode labels, label management is costly and complex.

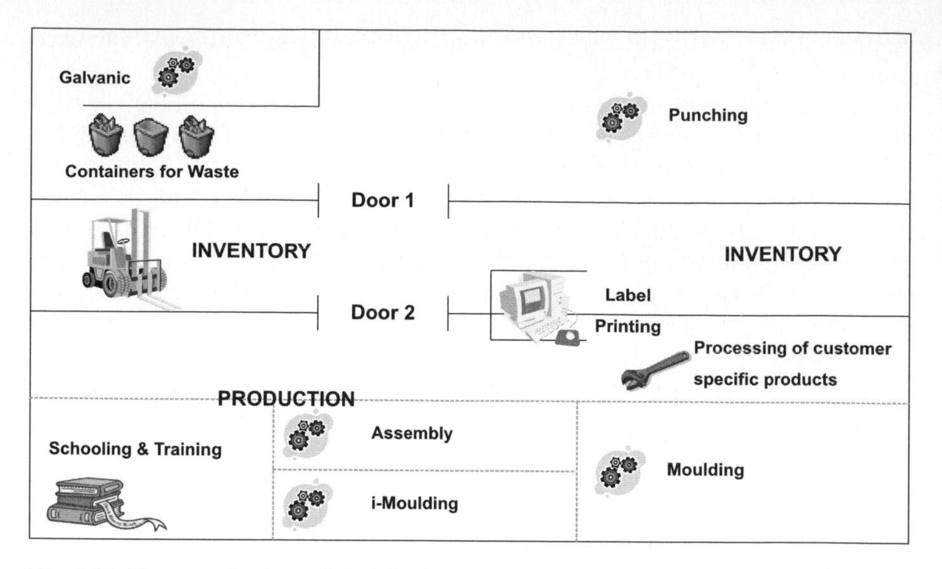

Fig. 4.16 The organization of CON's plant

Barcode labels are printed at a central location in the plant (see Fig. 4.16). After printing, workers send the barcode labels to the packing stations. After the containers return to the plant, workers must remove the labels on empty reusable containers in a cleaning process.

4.5.2 RFID Perspectives

Until now, CON has not employed RFID for any production task. Yet some processes hold the potential for improvements through the use of RFID. This section outlines possibilities for applying this technology and the benefits that may result. Among the potential benefits of RFID, we consider five possibilities:

1. Synchronization of processes with data tracks
2. Maintenance of process data
3. Container management
4. Uniform labeling
5. Process safety

We discuss these potential improvements and illustrate possible effects on the current production processes.

4.5.2.1 RFID Application Scenario #1: Synchronization of Processes with Data Tracks

Each production step must be booked into the back-end system after it is finished. This is done in order to keep track of all production processes and to ensure traceability. Before a production step can be started, the required material must be transported to the inbound of this step, and the previous production step must be booked. Because transporting and booking are independent, it is not ensured that both processes are conducted. According to CON's employees, sometimes no booking takes place even though the product has been forwarded to the next production step.

Here, RFID could possibly be used to improve process efficiency by ensuring that booking and transporting occurs in a synchronized manner. For this purpose, the internal transport units must be equipped with RFID tags. They could be put onto the coils for the copper straps and on the boxes for the plastic parts. This would allow direct coupling of transportation with the booking process. Readers could detect that an internal transport unit has been moved to the next production step.

Automatic inference on read events could be performed depending on the number of readers and the kind of information stored on the tags. In cases of incomplete information, at least missing bookings could be detected immediately, and the responsible employee could be informed instantly. Even the whole booking process could be triggered automatically by read events if more information were available on the tags. This would make transporting and booking synchronous and ease the maintenance of process data.

Two options exist regarding the installation of RFID readers at the plant floor. One option is to use only a few readers placed at intersection points of internal routes for material transportation (for example, at doors of the different plant halls; see Fig. 4.16.). The installation points must be chosen so that materials pass at least one reader when they are transported between operations.

This setup could be used to automatically check for missing bookings. An alarm could be triggered if a reader detects material movements that cannot be associated with a corresponding booking. Therefore, coils and boxes must be equipped with RFID tags that hold at least an ID. Using this ID and information about the production plans, software in the back-end system could infer whether bookings were left out and trigger alarms as needed.

From a technological point of view, the outlined application scenario is possible. But shielding of radio signals must be considered because of metallic products. Tracked objects are moved around in packed bulks. Consequently, problems due to shielding of radio signals may occur, at least for metallic products. This problem could possibly be avoided by transporting the products in a loosely packed manner (such as by using only every third position on wagons, as the one shown in Fig. 4.15). However, an advantage is that the products are moved in relatively small bulks, so collisions are therefore not a communication concern.

Another option is to place readers at each operation. This would allow for triggering the booking process automatically. Readers would detect a disappearing event of internal transfer units (coil or box) at the outbound of an operation. This would indicate that the operation was completed, and booking could be done automatically.

RFID tags with a unique ID on internal transfer units are sufficient for this application scenario. Here, readers would be required at every processing point. Compared with the scenario with few readers, this would require a higher investment cost for readers. On the other hand, working hours would be saved because the booking process could be automated. A technological advantage is that readers at every processing point could scan the tags individually rather than in bulks. This would ease finding installations where the reader signals are not shielded by metal parts.

A benefit in both application scenarios is that tagged objects are reused, i.e., it is a "closed-loop" application. Consequently, RFID tags would be reused, and tag prices of several euros would not be an issue. Even active tags are available in this price category, which would make communication more reliable.

4.5.2.2 RFID Application Scenario #2: Maintenance of Process Data

According to estimates of CON employees, a worker on the plant floor spends 15% of his or her working hours maintaining process data. Up to a third of this time is spent maintaining tickets for internal transportation. Parts of the data maintenance could be automated if tickets were partially, if not totally, replaced by writable RFID tags. This would increase worker productivity as well as reduce errors in manual data entries.

Having boxes and coils equipped with RFID tags allows for automatic association of the internal transfer units with metadata about their content. Thereby, manual writing of process information on paper sheets could be avoided or at least reduced. In general, two options exist for automatically storing metadata. One option is to write the data to an RFID tag on the internal transfer unit. In this case, a rewritable tag with a sufficient amount of memory is needed. Alternatively, the information could be written to a database in the back-end system. This scenario requires only internal transfer units to be equipped with an RFID tag that holds an ID. The tag ID could automatically be read and used as a key in the database. RFID would thereby allow metadata about processes to automatically be associated with the corresponding internal transfer unit. However, using a database instead of writing data to RFID tags would require investments into the back-end system. Careful analysis would be needed to determine which option for storing data would be more cost effective.

For the realization of this scenario, the data flow between process steps must be evaluated in detail. Information that is needed only by machines could be transferred via RFID tags and not appear on paper. However, workers require information to find the right material for the next step. This information must consequently be visible to workers. Two solutions can be considered to handle this information. One is to equip workers with mobile RFID readers. For this solution, the readers' radio frequency field must be reasonably limited in range to allow for easy identification of which object corresponds to a read event. An alternative is to employ a hybrid solution in which some information remains on paper.

4.5.2.3 RFID Application Scenario #3: Container Management

Numerous types of containers for shipping are stored at the investigated production plant. Different customers require different packing units for shipping, making container management a challenging task. As described above, containers can be at various locations at CON. Today, no detailed tracking of containers takes place at the production plant.

Automatic tracking of containers could be realized with RFID use. If containers are equipped with RFID tags, readers at different locations could update a database that holds the container's position. Alternatively, mobile readers could be used for a quick update of the inventory. This would reduce time for searching for containers, lower the risk of losing containers, and reduce the costs of renting containers by decreasing the required safety stocks.

For tagging containers, low-cost RFID Gen 2 tags could be used. These tags usually have 96 bits of storage, which is sufficient to uniquely identify a container. If these tags are used only internally, CON could label containers at the inbound. At the outbound, the tags may have to be removed again. However, customers of CON could possibly benefit from having RFID tags on these containers as well. In this case, CON may form coalitions for using the tags and for sharing their cost. Also, the coalition partners may agree on what data are to be stored on the tags to increase the efficiency of container management across company boundaries.

4.5.2.4 RFID Application Scenario #4: Uniform Labeling

Labels on the outbound shipping units vary from customer to customer. Labels differ in the paper being used and the information printed on them. Currently, CON handles this situation by having a central printing station where different labels can be printed on different papers.

The numerous formats could be handled much more easily if customers agreed to replace paper labels with RFID tags. Different information required on the various labels could be written on the same type of standard tags. For instance, RFID Gen 2 tags with user memory may be used. These tags can be rewritten with standard readers/writers. If these readers/writers are already available at packing locations, labels must no longer be created at the central location but could be written at the point where they are needed.

4.5.2.5 RFID Application Scenario #5: Process Safety

The accuracy of processes at CON currently depends on employees' attention and care during the process steps. For instance, workers must be careful not to accidentally mix up input materials for a process step or not to confuse labels. If RFID were used in the respective process steps, accuracy could be automatically inferred based on read events and ensured by alerts.

RFID readers at every step could be used to identify the materials that are about to be processed. Therefore, internal transfer units must be equipped with RFID tags. Metadata about the carried materials could either be stored on the tag or directly written into the back-end system, although the latter option would cause additional communication overhead and investments in database technology.

Data about machine configuration could be retrieved from the machine or the corresponding terminal PC. These data and information about the planned production tasks could be used for consistency checks to ensure correct machine configurations.

4.5.3 Costs and Benefits

The use of RFID offers a number of potential benefits for CON. Each application scenario would involve different investment costs and resulting benefits. Not all of the benefits are quantifiable in terms of cost savings. In the following, the major investments and savings for the different application scenarios are outlined.

Fixed investment costs include those for readers, RFID tags, staff training, and configuration of the software system. Tag costs are fixed because the tags cycle in a closed loop. Variable costs are those for software maintenance and replacements of defective readers and tags. Different requirements for the RFID tags' capabilities exist in the different application scenarios. Thus, tag prices may range from about 20 cents to several euros. Reader prices range from a few hundred to several thousand euros. Investment costs vary substantially regarding the benefits described above. On the one hand, the various scenarios imply different process costs (costs to adapt and redesign business processes). On the other hand, the number of required scan points varies greatly. The same applies to the savings that may be gained within the different applications.

The scenario for synchronizing processes with data tracks presents two types of cost savings. One is that labor time, required to reconstruct data for missing bookings, can be reduced. Fixing missing bookings can take up to 30 minutes of employee time, according to interviews with the staff. Furthermore, missing bookings can interrupt the production process. The frequency of such incidents and the value generated in the running production process determine the monetary benefit of avoiding these interruptions.

The application scenario for the maintenance of process data accounts for savings in labor time. Employees on the plant floor spend about 5% of their time copying data from and to accompanying tickets. This time can be saved if the data were transferred automatically via RFID.

In the container management scenario, RFID can reduce the overall costs for purchasing and renting containers. Tracing containers would allow reduction of the safety stock for containers, and fewer containers would need to be rented. Furthermore, loss of containers could be reduced, or external partners could be held accountable for losses at their sites.

In the scenario that proposes using RFID for uniform labeling of external shipments, labor time and hardware costs can be saved. This scenario is highly interlinked with other RFID applications. If RFID readers are in place, they can be used to write customer-specific information on tags attached to outbound material. In this case, expensive specialized printers for labels would no longer be necessary. This would result in hardware cost savings and increased productivity.

In the process safety scenario, production waste could be reduced by avoiding false machine settings. The resulting savings with this application scenario depend on the value of the wasted material and the cost of processing it.

4.5.4 Summary

This case study presents an analysis of the potentials of adopting RFID at CON. The analysis revealed five areas that hold potential for improvements at CON's plant. Theoretically, all of these scenarios are technically feasible. The company now needs to look at the details of the various options.

In CON's case, we would recommend focusing on closed-loop scenarios within the plant in the short term. This is because CON controls all factors of the setup and can gain experience before negotiating with partners about collaboration. Closed-loop scenarios would address the automatic synchronization of processes with corresponding data tracks, maintenance of process data, and process safety.

4.6 PAC: Aluminum Foils for Packaging

PAC is a manufacturer of aluminum foils; an example is shown in Fig. 4.17. PAC's customers mainly come from the food, pharmaceutical, and beauty industries. PAC conducted the case study with us as part of an investigation of whether and how RFID could be used to improve its production. Two applications accounted mainly for the initial motivation for considering RFID adoption. One application is to establish an emergency system that could be used in

Fig. 4.17 Example aluminum foil from PAC

case of back-end failures. For this application, the ability of RFID tags to store data and thereby decouple data management from the back-end IT system is of interest. The other application of interest is to improve traceability of products during the production. Here, the easy readout of RFID tags without line of sight and RFID's robustness in dirty environments are the main drivers for considering this technology.

4.6.1 Current Situation

In this section, we describe the current situation at PAC in terms of its IT infrastructure and production processes, providing a general overview of the manufacturing activities with a focus on aspects that impact the potential introduction of RFID. Here, we distinguish between two phases of production at PAC. In phase 1, aluminum foils are milled into a desired thickness and cut into a specific length and width. We refer to this phase as "preprocessing". In phase 2, the foils are refined. They are colored and/or coated with cellulose films. But on a high level, the production processes in both phases follow the same general pattern, which is depicted in Fig. 4.18.

In both phases, each production step begins by fetching the required materials from stock or material buffers on the plant floor. Subsequently, the machine being used is configured, and the processing starts. Between all operations, aluminum foils are reeled on carrier rolls that are transported on carrier frames. During the processing itself, machines wind off the input foils from the source roll, process the foil, and reel the foil back to a new target roll. In a cutting process, a foil from one roll can be split up into several foils and reeled onto 'child' rolls. Also, multiple foils can be reeled in several layers of one roll. This material flow must be tracked during the whole production process. Thus, tracking information and information about the conducted production steps is written to documents that accompany the carrier rolls or is written into the MES. Subsequently, the rolls are transported to material buffers, which are located be-

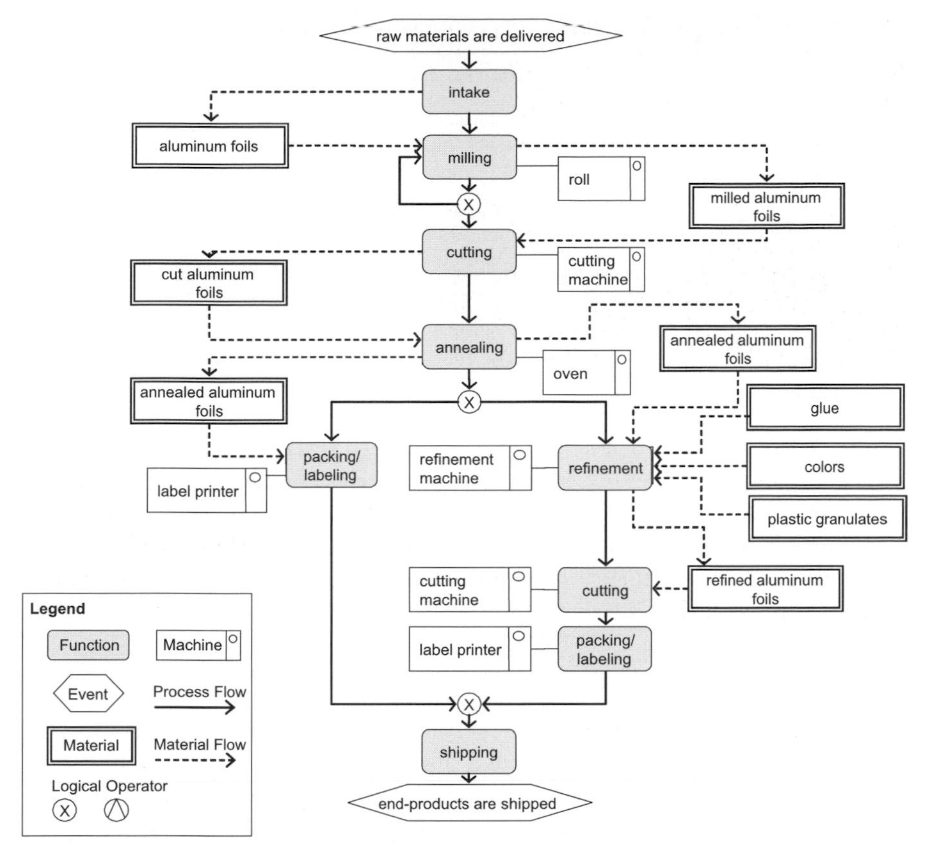

Fig. 4.18 General model of the production processes at PAC

tween the machines on the plant floor. Keeping track of these material buffers is the responsibility of the executive producer and is not supported by the MES or the ERP system.

PAC uses various methods and media for different aspects of the data management (see Table 4.3). Currently, the ERP system SAP R/3 and the MES HYDRA are employed for managing all data. Coarse-grained planning of the production process is done in the ERP system, while fine-grained plans are created in the MES. Part of the planning in the ERP system is the acquisition and management of required materials. If all materials are available, the production task is forwarded to the MES. Here, the concrete order of production tasks is determined and optimized. For instance, tasks that require the same raw materials are grouped together.

Documentation and control of activities in the production are also done with the MES. Terminals are situated at the machines on the plant floor for interaction with the MES. Via these terminals, employees select production tasks, configure machines, and enter information about the conducted activities. PAC

Table 4.3 Data maintenance conducted in all production phases

Data management	Location of operation	Information	Use
SAP R/3 (ERP)	Back-end system	Order information	Coarse-grained planning of the production
		Archiving of production data	Documentation of production and recalls
HYDRA (MES)	Back-end system, terminal PCs	Fine-grained production plans, machine settings, process data	Fine-grained planning of the production, control of the production process
Wax identification numbers	Foils	Order number and charge number	Identification of foils, association with accompanying tickets
Error reports	Roll carriers	Order number and charge number	Identification of foils, association with accompanying tickets, handout to the clients
Executive producers	–	Position of rolls in material buffers	Organization on the plant floor

keeps all data that are recorded in the production. This is necessary because customers demand traceability of each production step, including information about the raw materials being used. Clients from the pharmaceuticals industry especially request very detailed data records, and there is a tendency for PAC's customers to demand more and more fine-grained data tracks. Therefore, PAC plans to extend its data collection in the future. Currently, the production data are stored in the MES for 30 days and archived in the ERP system afterward. The MES holds data for about 3,000 production tasks at the same time, whereas the data volume in the ERP systems amounts to a few terabytes.

In addition to the data management in the MES and ERP system, PAC uses handwritten notes for documenting the production. Information about the conducted production step is written on paper documents or is entered into the MES. For instance, necessary interventions or machine problems can influence the products' quality and are documented. In cases of major quality problems, foils are removed from the process, and an error report is issued. Error reports and accompanying tickets are attached to the carrier rolls and are thereby associated with the corresponding foil. In addition, the documents are associated with the foils by means of special identification numbers. Workers also use these numbers for identifying the foils on the plant floor. The numbers are written with a wax pen on the last meters of the foil, which are removed when the material is loaded to machines. Therefore, the number must be rewritten after most operations.

These data management activities are conducted during preprocessing (phase 1) as well as during refinement (phase 2). Table 4.4 shows an overview

Table 4.4 Information carriers in the preprocessing phase at PAC

Storage medium	Storage location	Information
Control tickets	Machine	Quality records
Accompanying tickets	Carrier frame	Completed tasks, errors, corresponding charge
Material records	Carrier fra me	Completed tasks (for SAP system)
Annealing plans	Carrier frame	Annealing time
Product reports	Packaging	Quality information

of documents that are used in both phases of production at PAC. In addition, some phase-specific documents are being used. In the following we describe both phases in detail, focusing on the conducted production steps and the related data management.

4.6.1.1 Phase 1: Preprocessing

In the first phase of production at PAC, aluminum foils are preprocessed. That is, foils are shaped into the desired length, width, and thickness. Preprocessing is conducted completely independently from the refinement steps in phase 2. Although most preprocessed foils are subsequently refined, some foils are directly shipped to customers without further refinement.

Figure 4.18 shows a generic process model that fits production in the first phase. Information carriers that are used in this phase are listed in Table 4.4. Coarse-grained planning at the level of weeks is done in the ERP system SAP R/3. At this point, availability of the required materials as well as carrier rolls and carrier frames for internal transportation is checked. Subsequently, orders are released to the MES, where fine-grained scheduling takes place. Note that concrete processing orders are reflected in the MES but are scheduled manually based on the responsible employee's experience.

Production tasks are displayed on terminals on the plant floor according to the schedule in the MES. At the beginning of production, information about the order is retrieved from the back-end system, and the required materials are booked. At this point, the quality of the raw material is manually documented on a material report. Additionally, an accompanying ticked is created. The accompanying ticket and the material report stay with the foil during the whole production process. Workers use these documents to check and control the processing tasks that need to be accomplished and to record information about the production. Recorded information is used for tracking steps and controlling product quality. That is, conducted operations are documented along with information such as process changes, environmental temperatures, and materials used. These recordings may later be used for recalls if problems are detected.

The tasks of an operation are managed from workstations at the corresponding machines. Here, workers collect all documents from the foils scheduled for processing. Data from the accompanying documents are synchronized with the MES via terminals at the workstations, and consistency checks are performed. After successful checks, the processing is enabled, and an according protocol is signed by the executive producer. The tasks that are subsequently performed at the machine are documented on control tickets kept at each machine. Those tickets are scanned on the subsequent day and archived for 7 years.

After a step is finished, the foil's accompanying and material report tickets are updated with the information about the conducted processing and attached to the carrier frames holding the corresponding carrier rolls. Note that several foils may be reeled in layers on the same carrier roll. In this case, it is crucial that the layers be documented correctly on the accompanying tickets. In other cases, a foil from one roll may be cut into several child foils that are subsequently reeled to different carrier rolls. New accompanying tickets and material records are created to document the child rolls, and the documents of the parent roll are discarded. Identification numbers for the child rolls are created by extending the number of the corresponding parent roll with a counter. It is thus possible to trace all rolls that were created from a certain roll and to narrow recalls if quality problems are detected.

The described actions are performed at all operations in the first phase of PAC's production. In detail, this phase comprises milling, cutting, annealing, and shipping. Some of these steps may be conducted repeatedly or in varying order. Annealing, however, is always the last step before shipping.

During milling, foils are milled to the desired thickness. This step may be conducted repeatedly to achieve the desired result. In this process, foils are wound off their carrier roll, processed, and reeled to a new carrier roll. Therefore, all accompanying documents must be moved to the new carrier as well. The handwritten identification number on the foils ends must also be rewritten because a few meters from the foil ends are removed when the milling machine is loaded.

During cutting, foils are cut to the desired length. Thus, material from one source roll is reeled to multiple child rolls. This relationship between source rolls and resulting child rolls must be documented to allow backtracking of the production process.

Annealing is the last step before shipment. The purpose of this treatment is to harden the aluminum. Batches of foils are placed in a special chamber that heats up to about 400°C. The time a roll has to spend in the chamber varies and can be up to several days. Therefore, an annealing plan that describes the parameters for the annealing process is attached to the carrier frame. A correct annealing time is necessary for ensuring high quality of the material. However, the time that foils spend in the chamber is tracked only on the batch level; a worker registers when a batch is completely loaded to the chamber or completely taken out again. Thus, the exact time that a certain roll spent in the chamber is unknown.

Another concern in tracking foils during the annealing process is the hostile conditions inside the chamber. The 400°C heat renders the use of any paper labels or documents infeasible. Thus, only the aluminum labels can accompany the foils during the annealing process, and all paper documents are separated from their corresponding foil. At this point, only the wax pen numbers and the aluminum labels identify the foils. Both identification means are unreliable, according to employees. Wax numbers can be illegible or smudged, and aluminum labels can fall off during the annealing. Such cases pose the risk that the accompanying paper documents become associated with the wrong foil after annealing.

The final step in phase 1 is packing the preprocessed foils. The rolls are packed in client-specific carrier frames or boxes. These transportation units cycle in the supply chain between the costumers and PAC. However, PAC does not keep track of the location of the transportation units in the ERP system and cannot trace which client holds which unit. Consequently, clients cannot be held liable for missing carrier frames, resulting in a chronic shortage of transportation units.

Carrier frames are labeled with barcodes and transported along a partly automated packing station. At this point, the barcodes are automatically scanned (or manually registered if automatic scanning fails). Boxes are labeled with alphanumeric numbers that are manually registered, but PAC plans to replace boxes with barcode-labeled carrier frames in the next year. In addition to the labels, customer-specific products reports are attached to the transportation units. Product reports include information such as the results of quality tests and the weight of the rolls. Transportation units are shipped to the customers along with this information. However, a majority of these preprocessed foils are shipped internally to phase 2 for refinement.

4.6.1.2 Phase 2: Refinement

The second production phase at PAC refines the preprocessed foils from the first phase. Foils are laminated, varnished, and combined in layers. Examples of end products in this phase are lids for dog food and coatings for toothpaste tubes.

As in the first phase of production at PAC, order-specific information is loaded from the ERP system to the MES. Here, the planning period for the detailed production schedule is 2 weeks in advance; that is, machine times are allocated 2 weeks before production starts in order to ensure sufficient time for providing the required input materials. Apart from that, the planning is done in the same way as in the first phase.

All information concerning an order and the corresponding rolls is stored in the MES. In contrast to the first phase, the whole documentation for the production is directly done in the MES. Thus, accompanying tickets and quality reports on paper are not used in the refinement phase. Information in the MES is associated with the corresponding foils via barcodes on the carrier rolls. It is important to note that these barcodes refer to foils in a certain state of production. Consequently, all foils get new numbers, and barcodes need to be replaced after each operation. In addition to the barcodes, identification numbers are written on the foils with a wax pen.

A major difference to the first phase is the variety of combinations for input materials. During refinement, aluminum foils, plastic granulates, glues, and other materials are processed to create the final product. Subsequent to the refinement, rolls are cut into smaller parts according to customer-specific requirements.

In the refinement phase, multiple processing tasks are conducted in parallel by the same machine. Once started, the machines process the entire foil, and interventions during the processing are generally not possible. Thus, the process keeps running even if quality problems are detected. In such cases, the affected parts of the foil are marked and cut out later. Furthermore, changes of the input materials are documented in the running process. For instance, glues or colors may need to be refilled during refinement. Tracking which parts of the foil are processed with which input materials allows narrowing of recalls if quality problems are detected later on.

Subsequent to the refinement steps, foils are cut and packed for shipment. Workers use information in the MES to determine which cutting size is planned for the foils. Cut parts of the foils are marked with duct tape that holds the corresponding order number and identifier of the source foil. Subsequently, client-specific barcode labels and reports about the conducted processing are printed. These documents are packed together with the finished product and shipped to the clients.

4.6.2 RFID Perspectives

RFID is currently not applied in PAC's manufacturing processes. During our case study we analyzed which potential improvements may be enabled by introducing RFID applications at PAC. We considered the following applications:

1. Tracking foils in the preprocessing phase
2. Managing production data in the preprocessing phase
3. Emergency system

In the following section, we discuss each application scenario along with technological issues that must be taken into account.

4.6.2.1 RFID Application Scenario #1: Tracking Foils in the Preprocessing Phase

Between operations in the preprocessing section, foils are moved to material buffers. Carrier frames with foils on carrier rolls are stored on the plant floor. At this point, only the responsible executive producers know the location of the foils, but reflecting the foils' positions in a tracking system would pose several advantages. One improvement is that the risk of losing foils would be reduced. If employees forget about carrier frames, the plant floor and the foils must be searched, and production is delayed. Such situations could be avoided with the help of an accurate tracking system.

In addition, a tracking system would enable additional insights into the production process. The annealing operations could especially be monitored in greater detail. Currently, the annealing time for foils is recorded only at batch level, so it cannot be determined how long a certain foil was exposed to the heat. RFID could ease the tracking of individual foils and thereby enable documentation of the annealing process with fine granularity. This would improve control and may help address quality issues related to this process.

RFID could be used for realizing highly automated tracking at PAC facilities. Several technological options can be considered for this application. For identifying the foils, RFID tags could be mounted either on the carrier frames or carrier rolls. These tags must store the foil identifiers. Because these identifiers change, tags must either be rewritable or be associated with changing numbers in the back-end system. A special challenge is that the used tags must be very resistant to high temperatures if they should be applied during annealing. Special casings can protect RFID inlays even in extremely hostile environments. However, it must be evaluated which protection is capable of enduring long-term exposure to heat.

Different reader installations can be used to capture the foils' positions on the plant floor. Stationary gates would be a suitable place to register foils as they enter and leave the annealing chamber. Mobile readers could be used to capture foils in material buffers on the plant floor. This would require employees to walk around the material buffers and update the inventory. Here, workers may use RFID for quick registration of foils at certain positions. The position for the identified rolls may either be entered manually or inferred from additional position tags on the plant floor.

4.6.2.2 RFID Application Scenario #2: Managing Production Data in the Preprocessing Phase

During preprocessing, significant data management tasks are conducted using paper documents. This poses a risk of read or write errors, especially in the case of handwritten data. Information on these tickets is used by the workers to identify the materials on the plant floor and to choose the appropriate machine settings. The manual data maintenance is a potential source of errors.

For instance, if multiple foils are reeled in layers on one roll, the workers determine the order of layers via the tickets. Mixing up these positions results in incorrect processing of the material. Another problem of the current data management is that the information is not directly coupled with the corresponding objects. The paper documents are removed from the carrier frames before processing and placed back afterward. At this point, documents can potentially be mixed up. Another risk is that documents can fall off—especially in the annealing process—and get lost or mixed up.

The described problems may be solved by an application that uses writable RFID tags for maintaining process data. In this application scenario, RFID tags could be permanently attached to the rolls. Information would thus be directly coupled with the corresponding object and could not get mixed up or lost. Additionally, this solution would allow for automating parts of data management. Data about the conducted operations could be copied electronically from the machines or from the back-end system to the RFID tags' memory, so errors in manual data management could be avoided. For instance, information about the order of multiple layers on a roll could be copied from the machine that reeled the foils onto the roll. This would require an integration of machine data and process information with the middleware that controls the RFID readers.

4.6.2.3 RFID Application Scenario #3: Emergency System

An emergency system realized with RFID was explicitly requested by PAC's IT staff. This issue is of high importance because production at the investigated plant must currently stop if there is a back-end failure. In the refinement phase, all information is managed in the back-end system. Thus, information on how the process should be conducted is missing if the corresponding databases are not accessible. Such incidents have occurred in the past and caused downtimes of several days. Consequently, the IT staff highly desires the decoupling of critical functionality from the back-end system and the ability to ensure at least partial operation if the back end fails again.

RFID can be used as an enabler for an emergency system that can operate without the back end. This is due to the ability of RFID tags to store data. In the desired application, rolls would be equipped with rewritable RFID tags. Using the memory on the tags, process information related to the carried foils can be stored with the rolls directly. Therefore, required information must no longer be queried in the back-end system but could rather be retrieved from the tags. For instance, if the back end fails, this approach would allow continued printing of the correct labels at the packing station.

To realize an emergency system that is able to operate without network connectivity and without the back-end system, the business logic must be shifted to the terminal computers mentioned above.

4.6.3 Costs and Benefits

The main driver for implementing RFID at PAC is to establish an emergency system. This is because the company has already suffered from downtimes of the back-end system, which resulted in production being stopped for several days. Realizing an emergency system with RFID would require writable RFID tags with a few kilobytes of memory as well as one reader per workstation. These investments must be compared to the probability of downtimes of the back-end system multiplied by the average costs related to such an incident.

Additionally, RFID may help reduce production errors and improve product quality. This could be achieved by automating the production-related data management and thereby avoiding errors in manual data maintenance. The frequency of errors and the related costs determine the potential savings for this application scenario. Furthermore, improvements in data accuracy may help PAC analyze its processes better and identify potentials for improvement. However, the costs and especially the benefits of this scenario are hard to quantify.

4.6.4 Summary

In this case study we presented an analysis of the potentials of adopting RFID at PAC. We identified three main scenarios concerning how RFID can improve PAC's operations. The first scenario is to realize an application for tracking foils. Here, the ability of RFID to be read out without a line of sight and the possibility to provide protective casings for RFID tags are the major arguments for considering RFID instead of barcodes. The second scenario targets the maintenance of production data; RFID could be used to ensure that production information is associated with the right object. In the third scenario, RFID would be used to realize an emergency system. This system should ensure at least partial operation of production even if the back-end system fails. Here, the possibility to write information onto the RFID tags is exploited ("data-on-tag"). This allows decentralization of the business logic and its decoupling from the back-end system.

4.7 Summary

In this chapter we described six case studies illustrating the use and potential of RFID in the manufacturing industry. In all case studies we first analyzed the current situation on the shop floor. Then we illustrated how the current production could be optimized by applying RFID technology in diverse scenarios. We concluded the case studies with a cost and benefits discussion for each portrayed RFID application.

Generally, the case studies show that RFID technology holds many promises for improving manufacturing processes, while also exhibiting new challenges. The automation of object identification processes through RFID can help increase efficiency by reducing scan times and manual work, reduce errors due to manual data entry and analysis, and improve product tracking and tracing. Detailed data tracks can help increase product quality and narrow the extent of necessary product recalls.

Compared with barcode technology, RFID does not require a line of sight for scanning, enables simultaneous batch scanning, does not require the technological effort for high-quality printing, and is more resistant to physical influences such as dirt and scratches. Avoiding problems related to unreadable barcodes may help reduce the number of returns and penalties and increase customer satisfaction, especially in supply chains operating according to the JIS paradigm.

Despite the high potential of RFID technology, a manufacturer must consider a number of issues before starting an implementation. For example, environmental conditions (such as heat, the presence of metal or water, and the plant layout) may impact the applicability of RFID. Furthermore, the effort involved in creating and maintaining an infrastructure of RFID readers, terminal computers, and communication networks as well as the costs for the tags must be weighed against the benefits. For example, tag costs will become significant for high-volume, low-value products that are individually tagged. Apart from the hardware, robust and scalable software is needed to handle the processing of RFID data streams. We will discuss this in more detail in Chap. 5.

Last but not least, some of the discussed application scenarios require intense synchronization between supplier and OEM. If barcodes are to be replaced in cross-company communications, the parties have to agree not only on protocol standards (such as EPC and Gen 2) but also on data formats and processes.

The case studies show that applying RFID technology in manufacturing strongly depends on the individual scenario, the physical conditions at the plant, the existing infrastructure and expertise, and potential synergies with other projects. Therefore, only a specific case study with local field tests will provide a qualified answer. But the case studies presented in this chapter can help identify general potentials in a given case and find a starting point for detailed investigations including field tests.

Chapter 5
Lessons Learned

Coauthored by Lenka Ivantysynova and Holger Ziekow

In this chapter we discuss key findings of the case studies presented in Chap. 4. In Sect. 5.1 we give an overview of the terminology that is relevant for discussing RFID applications in manufacturing, and in Sect. 5.2 we present a reference model for production that we have derived from the case studies. This model captures typical activities on the plant floor and the corresponding data management issues. The description reflects an idealized process and focuses on aspects that appeared relevant for an RFID application. However, process steps are not always performed in an optimal manner in the real world. In Sect. 5.3 we present typical use cases of RFID that we identified in the case studies. Following this we discuss the trade-off between barcode and RFID technology in Sect. 5.4 and provide guidance for assessing cost and benefits in Sect. 5.5. In Sect. 5.6 we present basic functionalities of RFID infrastructures. In the Subsesquent section 5.7 we discuss functional and non functional requirements for IT infrastructures to integrate RFID in manufacturing. Along with an overview of upcoming paradigms for data processing that are useful in the context of RFID and manufacturing. Finally, we discuss hardware issues in the plant floor in Sect. 5.9.

5.1 Terminology

In this section we give an overview of the terminology relevant for discussing RFID applications in manufacturing. Devices such as *sensor gates, RFID tags* and *RFID readers* are used to collect data from the shop floor. They pass this data on to *terminal computers* or *edge servers* that are directly connected with the RFID readers and sensor gates. *Terminal computers* can host device controllers, which conduct low-level filtering of RFID read events. These computers are situated on the plant floor and are connected to devices like RFID readers or processing machines. For instance, a terminal to the *MES* may serve as a

O. Günther, W. Kletti, U. Kubach, *RFID in Manufacturing*
DOI: 10.1007/978-3-540-76454-0,

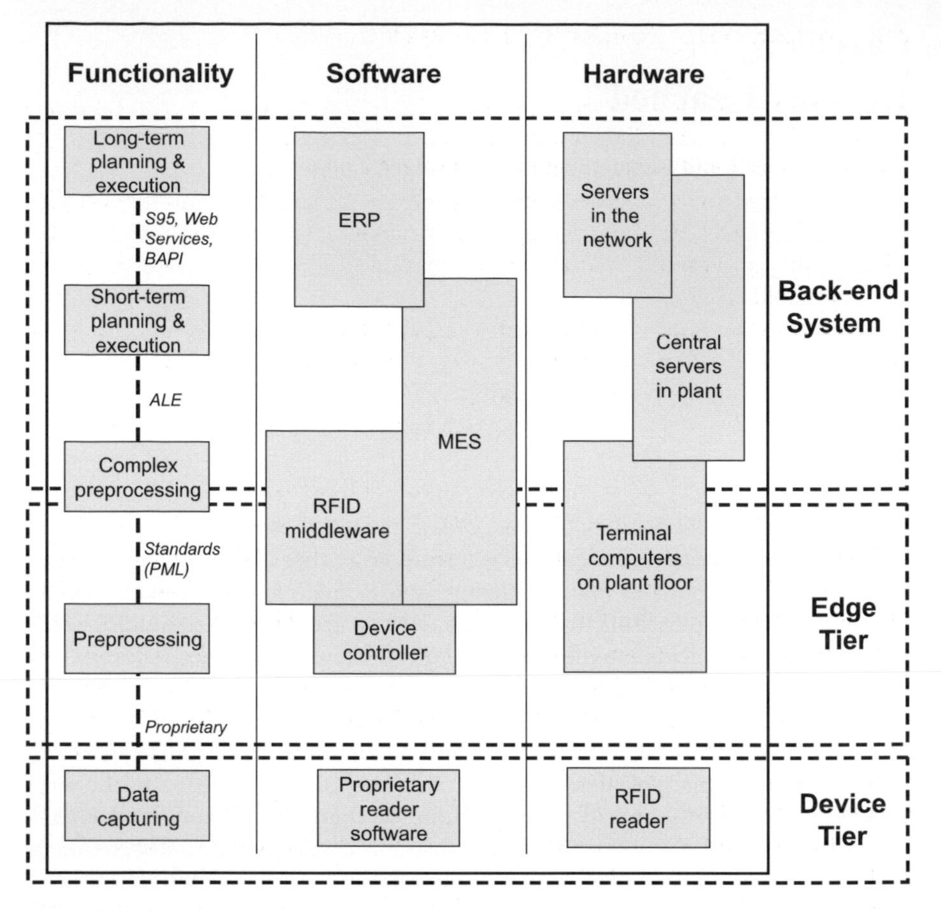

Fig. 5.1 System overview

terminal computer. More advanced data processing can be performed by *RFID middleware* in the *back-end system*. The back-end system comprises the *MES* and the *ERP* system. Figure 5.1 gives an overview of relevant functionalities and hardware and software components. The vertical order of the components implies which functionality is commonly realized in which software system and on which hardware the software is running. Note that the implicit association between functionalities, software components, and hardware in this figure is not fixed. Though components on the same vertical layer are associated in typical setups, different mappings between functionality, software, and hardware are possible, if not likely.

5.2 A Reference Model for Production

This section presents a reference model for production processes and management of the corresponding production data. The model focuses on activities that may be affected by the introduction of RFID. We derived this model from analysis of the case studies presented in Chap. 4. Although the production processes differed significantly among the companies we investigated, it was possible to identify common patterns in manufacturing and the associated information management.

A production process consists of a sequence of operations. An operation can generally be subdivided into the following 10 activities: selection, material fetching, tool exchange, machine configuration, consistency checks, processing, documenting, booking, loading into transportation units, and transporting (see Fig. 5.2).

The first step is to *select the correct operation* from the routing (e.g., bill of operations). The second activity is to *fetch the needed materials* for the operation. These materials must be retrieved from stock or a material input buffer near the resource and loaded into the machine. Materials are usually identified via the transportation units in which they are packed. However, in some cases, materials may also be marked directly with an identifier. When the materials are ready to be used, the machines need to be configured. *Configuring a machine* may include *mounting special tools to it*. These tools must be retrieved from the tools inventory. Because of quality issues and tracking, it may be required to maintain a history of all utilized tools.

Consistency checks may be conducted before *processing* starts. Such checks verify that the correct materials are used and that the materials have passed all required previous operations. When processing is completed, the conducted work is *documented*. Recorded information may include machine settings and reports about production errors. Thereby a track of the production is kept. It can later be used for process analysis and for help in narrowing recalls. One challenge in tracking is that materials may be *packed into transportation units*. This must be taken into account if materials are identified by their transportation units. Another challenge is that materials may be split up into several parts or combined into one part during assembly. This impacts how data tracks need to be recorded and retrieved.

An operation usually ends with *booking* the finished tasks into the back-end system (typically the MES). The data may be required for consistency checks in the following operations. Then the processed parts or work in process (WIP) are *transported* to the next operation, the stock, or a material buffer on the plant floor. Information about the materials' locations may need to be recorded in an inventory list or may be derived from the materials' current status in the production.

After production is completed, the finished products are transferred to a shipping area, where they are packed and labeled for shipment. A special chal-

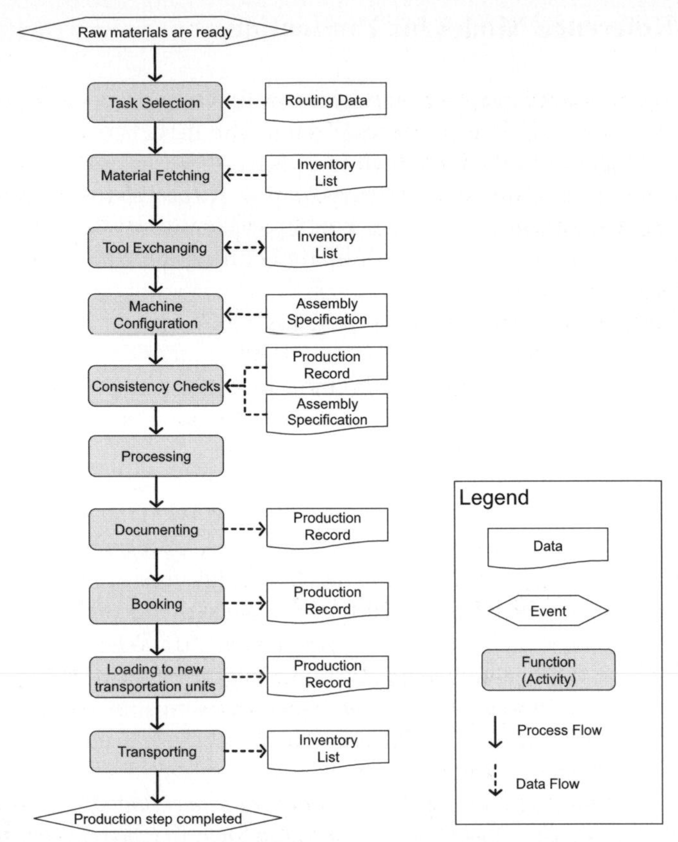

Fig. 5.2 Reference model for production

lenge is that different customers may demand different labels and different information on the labels. This may comprise different numbering and coding schemes for identification and product-related data.

Note that a particular production process may implement only a subset of the described activities. For instance, in many cases it may not be necessary to exchange tools on the machines. Thus, apart from the processing itself, each activity is optional in the generic reference model. That is, not all actions necessarily occur in every case, and others may be added. Still, the reference model presented above may serve as a general pattern that approximates most production processes. All activities in the derived general reference model may be supported by RFID or can influence RFID adoption. A software system supporting RFID in manufacturing should certainly take the steps of this reference model into account.

5.3 Motives for Using RFID

Based on the experiences drawn from the case studies, we now describe several general use cases for RFID. They typically are either a replacement of barcode technology or an application that can *only* be realized using RFID. We found that most of the relevant issues came up repeatedly in the case studies. The most common motivations for planning an RFID introduction were the following:

- Accelerating scan processes
- Extending scan processes for quality and efficiency
- Extending scan processes for narrowing recalls
- Reducing paper-based data management
- Automating asset tracking
- Reducing back-end interactions
- Unifying labels

5.3.1 Accelerating Scan Processes

Currently, many companies monitor their production processes by scanning barcodes or manually registering objects at certain check points. These approaches may have several drawbacks, depending on the particular process. Manual data recording is generally error prone and time consuming. Scanning barcodes can be automated in some cases but must be conducted manually if a line of sight cannot be ensured. Thus, scanning barcodes may also require time-consuming manual intervention. Depending on the particular process, manual scanning may account for a significant proportion of the employees' workload (Ivantysynova and Ziekow 2007). Consequently, manufacturers aim to reduce the time for manual scans or to automate the process.

One important property of RFID technology is the ability to automatically read information from the tag without a line of sight. This property can significantly speed up the scanning of identifiers because of the following three advantages of RFID technology: First, some objects in the manufacturing plant are shaped in a way that barcodes must be applied in places that are difficult to scan (see Sect. 4.1, 4.5). In such cases, RFID can help speed up manual scan processes (García et al. 2003). This property may also allow the automation of manual scan processes—a second major advantage of RFID technology. The third advantage occurs when whole batches of objects must be identified (see Sect. 4.1, 4.2). With RFID, all objects in these batches can be captured at once. Scanning without a line of sight and the capture of whole batches of objects reduces the need for manual interaction (Chappell et al. 2003). In cases in which line of sight for traditional identification methods is hard to achieve, RFID reduces the time that workers spend on scanning. This in turn enables businesses to cut labor costs and improve productivity.

5.3.2 Extending Scan Processes for Quality and Efficiency

Manufacturers have a high interest in getting more insight into the operations on the plant floor. Analyzing information about activities in production and measurements of environmental conditions can help identify causes of quality problems and point out potentials for improvements. Furthermore, analyzing live data can enable fast reactions to exceptions in the process. In general, a larger data set allows for better insight into the process. Therefore, many companies aim to extend data recording on the plant floor.

With RFID technology, production activities are typically much easier to monitor in detail. As mentioned above, the capability of scanning without requiring a line of sight enables automatic identification in more situations than barcodes do. Thus, with the help of RFID technology, new scan points can be introduced without increasing the staff's workload. Furthermore, RFID labels are typically more robust to hostile conditions that may occur in the production environment. For example, barcodes can become unreadable because of exposure to dirt, heat, or mechanical influence. RFID tags have the advantage that they can be covered in protective casings (DeJong 1998). Thus, in some production environments, RFID tags may work more reliably than barcodes (see Sects. 4.1, 4.4, 4.6). On the other hand, there are environments in which barcodes still beat RFID by a margin; this is especially true if a lot of metal is present. One needs to look at each particular case in order to make a well-founded recommendation.

Some of the investigated companies have stringent requirements for process reliability and process documentation (e.g., consistency checks to ensure that no process step is skipped). In one example from the case studies, the investigated company scanned the WIP parts after every operation to be compliant with customers' demands (Sect. 4.1). RFID can help meet such demands by simplifying the process of scanning items; thereby the technology also makes it easier to comply with regulations for process documentation.

Exact information about which object was manufactured out of which components and materials is required in order to identify all products that potentially include flawed parts. In addition, fine-grained and reliable data records can be important in legal disputes (Sect. 4.1). A company may be sued if malfunctioning products cause damage. In this case, data records are important to prove that the production was conducted in a way consistent with the state of the art. Here, sensor data can help detect the cause of failure and further narrow the scope of potentially affected products. Evaluating this data can provide additional insights about performance measures such as cycle times. It can even help identify the cause of quality changes.

RFID technology can also be used to ensure accurate and real-time reporting about the production status. Typically, a production step is reported back to the MES after processing is completed. This information may also be required in future consistency checks and for later process analysis. As the case studies have shown, it is not uncommon that manual reporting is sometimes forgotten or not conducted in a timely manner (Sects. 4.5, 4.2). Occasionally, production process information is even reported in the wrong order. Possible consequences are inaccurate data tracks, wrong status information, or even interruptions in the production process. These problems may be overcome if RFID tags are applied to the materials or to the transportation units. In such setups, RFID readers could automatically detect whether materials are transported to the next process step. Consistency would thereby be guaranteed.

5.3.3 Extending Scan Processes for Narrowing Recalls

Narrowing the scope of recalls was a major concern for many of the investigated plants. Manufacturers may have to pay high penalties for each object that is called back. Recalls can significantly reduce companies' revenue and their reputation with customers. It is therefore a major concern to limit recalls and their effects to the largest degree possible.

To narrow recalls, it is first necessary to locate the cause of failure as specifically as possible. Commonly, a faulty material or malfunctioning machine is identified.

In the second step, all potentially affected products must be located. This requires a detailed track about which materials and intermediate products are assembled to which finalized products, as well as information about these products' location. If the cause of failure was a malfunctioning machine, it is important to know the exact period of time when the machine did not work properly and which products were processed during this period.

To limit recalls to the maximum extent, companies aim to create fine-grained tracks of their internal material flow. However, manual tracking or tracking by using barcode technologies may not always be feasible. Time-consuming scans, the inability to create a line of sight, or hostile conditions can limit the number of possible checks and therefore the granularity of data tracks.

As mentioned above, RFID can help introduce additional scan points in the production processes. This is partly because RFID tags can automatically be read out in more situations than barcode labels can. Additionally, RFID tags can handle hostile conditions such as exposure to dirt. Furthermore, RFID tags with sensing capabilities can help to determine the cause of failure and to rate the damage. All these properties of RFID render the technology helpful in narrowing recalls where more traditional approaches fail (Sects. 4.2, 4.6).

5.3.4 Reducing Paper-Based Data Management

Paper documents that accompany the WIP are currently a common method for recording and maintaining data throughout the production process. These paper documents are transported along with their corresponding materials and are used to record data about the production process. Additionally, these documents can hold information about how to conduct subsequent operations. The accompanying documents are usually only loosely coupled with the objects they belong to. That is, documents move along with the corresponding objects but are physically separated from them while data is written on the paper. This may cause a mix-up of documents and incorrect data maintenance. Furthermore, handwritten notes occasionally cannot be deciphered, which effectively constitutes an information loss.

Correct data tracking of the production is crucial in many cases. Customers may demand high-quality data tracks, or the company itself may need those tracks for recalls and legal disputes about liability. Furthermore, the data are used for steering and controlling the production itself. Thus, mixed-up documents can lead to production errors and loss. Consequently, companies seek a way to ensure that data are recorded correctly and are permanently associated with the right object.

RFID tags with writable memory can be used to store data from accompanying documents right at the corresponding object. This way, information cannot get lost or accidentally mixed up. This solution may also be leveraged to automate some of the manual data maintenance. For instance, records of the conducted operations can automatically be written from machines to tags. Automatic data maintenance would account for time savings and a reduction of errors.

Alternatively, information about an object can be stored in the back-end system. In these cases, an identifier for the corresponding object is needed. Depending on the particularities in the targeted application environment, barcodes can be a suitable alternative to RFID. As mentioned, however, the application of barcodes may not always be feasible, such as in dirty environments. Furthermore, applying barcodes may reduce the degree of automation in cases where a line of sight cannot be created automatically.

Other trade-offs concern the required adaptations of the back-end system. In order to guarantee fast response times and high availability, an appropriate network infrastructure and software system for the back-end is required. The investment in such an infrastructure must be traded against the investment in RFID tags and readers, and the resulting network load must be considered carefully. However, even when data are stored locally at an RFID tag, the back-end system typically keeps a copy of this information, a situation that may cause problems regarding data synchronization.

5.3.5 *Automating Asset Tracking*

Knowing the spatial location of assets can be crucial for ensuring the production processes. All materials for an operation and the required tools need to be at the designated machine in time. If timely fetching of required materials cannot be ensured, the production process may be interrupted and result in reduced productivity.

In general, companies follow two strategies to avoid downtimes due to missing assets. One strategy is to fetch the required materials and tools with a long time buffer before production starts. This allows time to react when the assets cannot be found and must be searched for (see Sect. 4.4). But in this approach it is necessary to schedule the production in the long term in order to have the work plan ready for asset fetching. Consequently, this strategy is not feasible for companies that seek high flexibility and reactivity in using their production lines.

The other strategy to reduce fetching times for assets is to keep a detailed track of their positions. This reduces search times and makes planning of the fetching time more reliable. But keeping an accurate and updated track of assets typically requires interferences (often manual). A special challenge is to track materials in material buffers on the plant floor. Those material buffers are often only loosely structured, and their content is changed dynamically. However, tracking objects in widespread stocks can be challenging and time consuming as well.

RFID technology has the potential to facilitate the recording of assets' positions. For instance, assets can be equipped with RFID tags and registered by stationary readers at key points. Alternatively, mobile readers can be used to quickly register items at certain locations. The RFID technology thereby leverages detailed asset tracking and can help avoid search times for materials and tools (Lampe and Strassner 2003).

5.3.6 Reducing Back-End Interactions

Some of the investigated companies expressed the desire to reduce interaction with the back-end IT system. In one case, the network infrastructure and the back-end database were perceived as unreliable (Sect. 4.6). Consequently, production IT systems should also work during temporary disconnections from the back-end servers. In another case, the company's network and back-end computers were hardly able to serve the demanded response time (Sect. 4.1). The company's IT staff predicted significant bottlenecks when data volumes increase in the future. In both cases, RFID tags with writable memory ("data-on-tag") could help decouple the processing of business logic from the back-end system and distribute the workload. Currently, interaction with databases in the back-end system is needed to retrieve task-related data at each operation.

With the help of RFID, the workload of the back-end system as well as communication with back-end databases can be reduced significantly. The capability of RFID tags to store up to several megabytes of data enables a novel distribution of data and business logic. Data needed for consistency checks and process control can be stored directly on the RFID tags of the corresponding objects. Usually, this comprises descriptions of the referring order and historic data about the conducted operations. For instance, consistency checks that verify whether all needed steps were conducted can be performed solely on the basis of the historic data stored on the tag.

Moreover, by using the memory on RFID tags, the business logic for checks and process control can be moved close to the point of operation. For instance, the business logic can be run on terminal computers connected to RFID readers and machines on the plant floor.

Higher-class RFID tags could enable more distribution of business logic. Programmable readers can perform check operations locally and even independently of terminal computers. Furthermore, smart sensor tags can process business logic on the items themselves. Early examples of such applications have been investigated, such as in the CoBIs project (Spiess 2005). This project funded by the European Union deals with distributing business logic to smart sensor nodes. Prototypes have been developed with partners that include BP and Infineon.

Moving the business logic closer to the point of action on the plant floor helps to ensure fast system responses without tuning the back-end databases and the network infrastructure. It also helps increase system reliability. This is because in a distributed system, device failures affect only small parts of the infrastructure.

5.3.7 Unifying Labels

The case studies confirmed that manufacturers face challenges in handling labels at the outbound shipment. Different customers typically demand different barcode solutions for labeling transportation units and packages (Sects. 4.1, 4.5). These differences can be in the demanded label format, coding scheme, and/or the information on the label. Therefore, manufacturers have to manage and print a wide range of different labels for their shipment processes. Another challenge is that customers may claim financial compensation from the manufacturer if barcode labels are unreadable. In the case study in Sect. 4.1, a customer's production line stops if a barcode is unreadable, and the resulting costs must be reimbursed by the manufacturer who printed the barcode. Consequently, a high effort to create reliable labels is in order.

The demand for high-quality labels that hold diverse information in various formats poses challenges in managing and printing barcode labels. Often, special printers are used to achieve the desired print quality and to handle different label formats. Because these printers are usually expensive, some manufacturers create labels at a central location to reduce the number of needed printers. Yet printing at a central location has the drawback that the labels must be transported to the respective packing stations. This causes extra time and bears the risk that labels could get mixed up.

RFID technology can be used for unified labels that abstract from the physical representation of data. Radio frequency protocols such as those defined in the ISO 18000 or EPCglobal standards provide well-defined ways to access data on RFID tags. For instance, the Gen 2 standard specifies an optional user memory that can be used for arbitrary purposes. With adherence to such standards, customer-specific information can be written on the same kind of label in a standardized way. This holds even if customers require individual coding schemes for the data.

Abstracting from the physical representation of data allows for using standard RFID readers to create customer-specific labels. Thus, one kind of reader could be used for all labels, and specialized printing stations would no longer be necessary. Note that RFID readers may be available at packing stations for scanning logistic applications. In contrast to barcode scanners, RFID readers also have the ability to write on tags, so only one device for both reading and writing is needed. To achieve this abstraction, though, the supply chain partners need to agree on application of RFID in general and on the frequency to be used in particular. Within a given radio frequency spectrum, the readers' software can support different communication protocols and thereby abstract from different tag versions.

Besides the advantages in unifying different label formats, RFID can help increase the reliability of labels. This is especially significant for environments where dirt or mechanical influence can affect the barcode. It is generally easier to protect RFID tags from mechanical damage such as scratches or loss. This is because RFID tags do not need to be applied on the outside of an object where they are visible. Instead, they can be built into the product during the manufacturing process, be applied inside the product's casing, or be protected by a special casing for the tags' inlay.

Note that the discussed cases for label handling affect operations not only at the manufacturers' site but also at their customers' sites (original equipment manufacturers, etc.) down the supply chain. Using RFID tags as uniform labels therefore implies close coordination across the whole supply chain (or at least significant parts of it). Customers would at least need to equip their intakes with RFID readers. Having incoming items labeled with RFID tags would leverage an extended use of this technology within the customers' processes and may lead to significant productivity gains there as well.

5.4 RFID Versus Barcode

In this section we discuss if and how the previously described use cases could also be realized using barcode technology. This comparison is an important part of the evaluating RFID projects because labels and readers for RFID require higher investments than corresponding barcode equipment does. Consequently, technical arguments or process requirements must account for the economic feasibility of an RFID adoption. Furthermore, it must be discussed which RFID tags work best for a certain use case. Here, technological properties of tags must be considered as well as standards, tag costs, and expected future developments.

Subsequently, RFID applications are discussed with regard to the use of active tags, passive tags, and barcode labels. Regarding passive tags, we further distinguish between tags that hold only an ID and tags that are equipped with additional user memory. We consider active tags to have an ID plus additional user memory, as this is usually the case.

5.4.1 Monitoring Processes

RFID applications for process monitoring exploit the property of RFID that no line of sight is needed. This allows for easy automatic reading of the tags. Thus, tags can be read without manual intervention, and the booking process can be automated.

Theoretically, the same degree of automation could be achieved with barcodes. But this would require modifications of the production process to ensure line-of-sight readouts after each step. This could be achieved by using conveyers for transportation, for instance. However, the need for such an investment often renders a barcode-based approach infeasible.

5.4.2 Management of Process Data

Process data management can often be further automated by using RFID. The idea is to automatically communicate process data between the operations, rather than using manually written papers. The data can be communicated on the tags or via the back-end system.

Use of the tag memory is possible only when the storage is large enough. This is the case for most active tags and for tags with user memory. If information is stored on the tags, investments in the back-end system could be avoided. Database investments would be required if tags are used that hold only an ID. Which option is most cost efficient depends on the tag prices and the cost of extending the back-end system.

Barcode labels can theoretically be used in the same way as RFID tags without user memory, but if process data are no longer manually written on paper, workers must identify internal transfer units by their labels. Unlike RFID, barcode labels would require time-consuming manual scanning with a line of sight.

5.4.3 Container Management

In use cases for container management, RFID technology could be applied to keep track of container movements. Scanning without line of sight allows for easy observation of container positions. Any RFID tag capable of holding an ID could be used for this application. Yet the tag price plays an important role, even though tags can possibly be reused. One reason for this is the relatively high number of containers; another reason is that containers leave the plant. Consequently, manufacturers cannot always ensure that containers are returned.

It is often attractive to use low-cost RFID tags for container tracking. Ultrahigh-frequency (UHF) Gen 2 tags may be an option. The Gen 2 standard is well established and was designed for logistic processes. Applying this standard would leverage use of these tags by customers and other supply chain partners and may allow sharing of the investment costs. Furthermore, tags for this standard are already being produced in large quantities and are relatively cheap.

Using barcodes as identifiers would also be an option. However, line of sight must be ensured for the scanning process. This would cause several manual interventions and additional cost for workers. Moreover, barcodes are less robust than RFID tags and may be rendered illegible or get lost. Thus, applying barcodes for container tracking may not be cost effective.

5.4.4 Uniform Labeling

The use case for uniform labeling targets the challenge of handling labels for the shipment process. Different customers require different information and different label formats for shipment. The different formats could be realized on the same kind of RFID label if the customers agreed on using RFID in the shipment process. Information for each customer can be encoded in the desired format and written to the RFID tags.

For this use case, tags with sufficient storage, such as active tags or passive tags with user memory, are required. Because labeling packing units with active RFID tags would be cost intensive, passive tags with user memory are often preferable.

5.4.5 Process Safety

Additional consistency checks are often added to the production steps for process safety. Information about the content of internal transfer units as well as configurations from terminals and the production planning are required to perform consistency checks.

Table 5.1 Applicability of radio frequency identification and barcode technologies

Use case (section where discussed)	Active tag	Passive tag		Barcode label
		Only ID	Memory	
5.4.1	Possible	Possible	Possible	Not possible
5.4.2	Possible	Possible with additional database	Possible	Possible with additional database and additional workload for manual scanning
5.4.3	Possible but cost intensive	Possible	Possible	Possible with additional workload for scanning
5.4.4	Possible but cost intensive	Not possible	Possible	Not possible
5.4.5	Possible	Possible with additional database	Possible	Possible with additional database and additional workload for manual scanning

At least an ID must be read to retrieve information about the internal transfer units. This could be done using passive ID tags without user memory or barcodes. In both cases, though, additional database lookups are necessary to retrieve the required information. If barcodes are used, manual intervention may be required to ensure line of sight for the read process. This may cause additional workload and thereby additional cost. In contrast, RFID tags with sufficient memory could enable automatic consistency checks without additional database queries or manual intervention.

The applicability of RFID and barcode technologies for these various use cases is summarized in Table 5.1.

5.5 A Guide to Assess Costs and Benefits[1]

In this section we discuss ways to compare costs and benefits of an RFID implementation. Based on the different motivations discussed above, we provide basic guidelines for how costs must be considered and how benefits can be estimated.

We discuss costs and benefits for each of the RFID applications previously described as well as costs that occur independently of the particular use cases.

[1] This section is based on "A guide to assess costs and benefits for RFID investments in manufacturing", by Ivantysynova, Ziekow, and Günther, ICIS-2007 6th Workshop on e-Business, Montréal, Québec, Canada, 2007

We outline how benefits can be quantified and which costs must be taken into account. We also provide basic equations that capture major factors for savings and investments. It is important to note that these equations can serve only as rough guidelines for a full-fledged calculation. Analyzing a specific company would require extending and modifying the basic patterns provided by our framework.

Some costs and benefits are very difficult to calculate, and some aspects may not be measured in monetary terms at all. Like most complex IT projects, on the one hand RFID introduction can incur hidden costs, especially costs for changing and adapting existing business processes. On the other hand, RFID may imply benefits beyond those that are easy to measure and quantify in monetary terms. Note that the implementation of an RFID system can be directly compared to any IT project: The IT project's costs of integration, support, training, and maintenance are much higher than the actual purchase price of needed hardware and software. For this reason, the costs should be calculated with a total cost of ownership (TCO) analysis (Wolf and Holm 1998). A complete TCO analysis spans a specific period of time (such as 5 years) and includes expected and estimates of unexpected costs specific for the company. Therefore, a complete TCO analysis cannot generally be done. Consequently, in this section we restrict the discussion to general aspects that can be measured in monetary terms and that apply to any manufacturer.

5.5.1 Costs

Even though implementation and running costs for an RFID application depend on the particularities of the specific use cases, some general cost factors are likely to occur in all cases. Chappell et al. (2003) have presented a schema of various components of a possible RFID rollout. These components include hardware costs such as tags and readers, installation, tuning, software, and maintenance costs. We describe in more detail which costs to consider and show against which benefits the costs should be traded. We also present a schema for analyzing running costs.

We consider closed-loop scenarios with reusable tags as well as use cases in which tags are used only once. The costs targeted in this section must be traded against the benefits described in Sect. 5.3. We discuss fixed costs at the beginning and variable costs at the end of this section. Fixed costs can be captured by the following equation:

$$C_{\mathrm{F}} = S + T + H_{\mathrm{RR}} + H_{\mathrm{NT}} + H_{\mathrm{RT}} + H_{\mathrm{TC}} + M + I \quad (5.1)$$

where

C_{F} Fixed costs
S Costs for software

T	Costs for training staff
H_{RR}	Costs for RFID readers
H_{NT}	Costs for network technology
H_{RT}	Costs for reusable RFID tags
H_{TC}	Costs for terminal computers
M	Average costs per hour of maintenance
I	Integration costs

S refers to the fixed cost for additional software that is needed. This may comprise device controllers for readers and middleware for integrating RFID data with the back-end system. Furthermore, additional components with the desired application logic may be needed. How much additional software is required depends on the capabilities of existing systems on the plant. For instance, an MES may already provide some functionality of the needed middleware.

T refers to training costs. If the local IT department is in charge of configuring and maintaining the new software, the responsible administrators need to be trained. Furthermore, staff on the plant floor interacting with RFID tags and readers must be introduced to the new technology.

H refers to the hardware cost for devices that need to be purchased to run the desired RFID applications. This may comprise costs for readers (H_{RR}), reusable RFID tags (H_{RT}), terminal computers (H_{TC}), and network technology (H_{NT}). The investments in reader hardware depend on the technology of choice and the total number of reader devices. Compared to setups with mobile readers, it is generally easier to achieve high read ranges and reliable capturing of RFID data if fixed devices are used. However, in some use cases, mobile readers may allow companies to drastically reduce the number of needed devices because they can subsequently be used at multiple locations (Sects. 4.4, 4.6). For instance, inventory can be managed by using stationary readers at each shelf in the storage area (smart shelf) or by having workers scan the shelves with mobile readers.

H_{NT} refers to the communication hardware needed for linking the RFID readers with the back-end system. Stationary readers may include network devices for communication via Ethernet or wireless LAN. Note that such network technology is already available on many plant floors due to other applications (such as for communication with terminal computers). In such cases, only little or no investments for extending the networks are necessary. The availability of wireless LAN is a requirement for many applications of mobile readers. Direct wireless linkage with the back-end system is necessary if the captured RFID data must be evaluated in near-real time. For applications without real-time constraints, mobile readers may require no additional network technology at all. In such applications, captured RFID data can be stored in the reader's buffer and synchronized with the back-end systems later on. For instance, after every shift workers can plug readers into a cradle that connects to the back end.

H_{RT} refers to the cost of tags that cycle within desired application. In many use cases of RFID in manufacturing, RFID tags either do not leave the plant at all or cycle between the manufacturer and its supply chain partners. In such cases, tag costs can be considered as fixed costs (apart from necessary replenishments of damaged tags). The investment cost for tags highly depends on the desired capabilities of the tags and the scale of the application. Tag prices can vary from a few cents to about 100 euros. RFID tags at the lower end of the cost spectrum are usually passive labels that can store an identifier of a few bytes in length. Tags with extended storage, active tags with sensing circuits, and tags with ad hoc communication capabilities mark the higher end of the cost spectrum.

H_{TC} refers to costs for additional PCs that may be needed on the plant floor. These PCs run software to filter and aggregate the incoming RFID data and forward it to higher-level system components. Note that in some cases existing PCs such as MES terminal computers could be used for this task. However, the amount of any additional processing power needed would depend on the computers' workload and the frequency of RFID read events.

M covers maintenance costs for the RFID application. I refers to integration cost in the introduction phase of RFID. This comprises necessary tests and consultancy in the planning phase of the application. Details of the desired use case must be planned, and suitable technological setup must be designed. This includes selecting from the various options for software and hardware configurations. The company must choose appropriate communication frequencies for RFID tags and suitable standards for the radio frequency protocol. Furthermore, the data flow through the various software modules in the back-end needs to be determined, and business processes must be linked to the incoming RFID events. Subsequent to this initial application design, the company needs to conduct hardware tests in the application environment. In general, the performance of tags and readers in a particular environment is hard to predict. For instance, the presence of metal can distort radio frequency communication. Additionally, the tagged items and their casings may shield the radio frequency signal. Such effects must carefully be tested before the application can be rolled out at full scale.

C_F refers to the fixed cost for the respective RFID application. Note that if there is an alternative solution without RFID (e.g., barcode), the cost difference for this alternative solution must be calculated:

$$C_{FA} = C_F - (S_A + T_A + H_A + I_A + M_A) \tag{5.2}$$

where

C_{FA}	Additional fixed costs for an RFID application
S_A	Software costs for alternative solution without RFID
T_A	Training costs for alternative solution without RFID
H_A	Hardware costs for alternative solution without RFID
I_A	Costs for the integration of RFID alternative
M_A	Maintenance cost of alternative

The term ($S_A + T_A + H_A + I_A + M_A$) in Eq. 5.2 represents the fixed costs of the given alternative solution. In many cases, RFID is used as a replacement for identifiers such as barcodes or alphanumeric numbers. If these identifiers could completely be replaced, the costs for the related infrastructure could be saved.

In addition to the discussed fixed costs, we need to add variable costs of an RFID application to the total cost calculation. Here, we have to consider two cases. In the first case, RFID tags are reused in a closed-loop scenario. These are use cases in which RFID tags do not remain on the product and are reused during production. In the second case, the RFID tags are used only once and are subsequently discarded or left with the customer.

In the first case, the tags are removed from the objects (the products). Here, the costs for reusing the tags come to the fore (Tellkamp and Quiede 2005). We can sum up the general cost factors in the following equation:

$$VC_C = T \cdot L \cdot (A + R + TR) \tag{5.3}$$

where

VC_C	Variable costs in closed-loop applications
T	Service life of the application
L	Number of RFID labels applied per hour
A	Costs for applying a label
R	Costs for removing a label
TR	Costs for transporting a label

T is the expected service life of the application in hours. (Note that we use hours as the time unit throughout this section.) L denotes how many items per hour are labeled with RFID tags in the application. If RFID tags are applied directly to a specific part of the product, the number of tags equals the number of manufactured products per hour. However, RFID tags can also be applied to several parts of the product or to material carriers that hold more than one object.

The variable A refers to the cost of applying an RFID tag to the target object. Note that these costs must be considered as cost per tagged item. If tags are applied manually, this translates into the corresponding labor cost. Considerable effort may be required in applications in which RFID tags are attached directly to the product parts that move through the production process. In many use cases, RFID tags could be applied to transportation units that cycle on the plant floor (such as material carriers). In such cases, tags are applied only once. However, the data written on the tag (or associated with the tag) must be changed in each cycle. Depending on the particular setup, this task may require manual intervention, which results in variable labor costs.

R refers to the costs per tagged item that occur if RFID tags are removed from the corresponding object at the end of the production process. This needs to be done in closed-loop scenarios where reusable tags are directly applied on parts of the products. Removing tags accounts for additional labor costs.

However, no removal is necessary if tags cycle on transportation units on the plant floor.

TR refers to the cost of transporting reusable RFID tags between the points of application and removal. Note that these costs must be considered as cost per tagged item in Eq. 5.3. Transportation is necessary for reusing tags that are applied directly to the product. Tags may just be transported within the plant floor in applications that are restricted to one plant, but advanced RFID applications may span between several production steps in the supply chain, and tags may need to be transported between different plants.

In cases in which the RFID tags are used only once, we need to adapt the calculation of variable costs as follows:

$$VC_N = T \cdot L \cdot (A + HT - CO) \tag{5.4}$$

where

VC_N	Variable cost in cases in which RFID tags are not reused
T	Service life of the application
L	Number of RFID labels applied per hour
A	Costs for applying a label
H_T	Hardware costs for one RFID tag
CO	Compensation payments

VC_N refers to the variable cost for RFID applications in which RFID tags remain on the product and, consequently, cannot be reused. The factor T is the expected service life of the application in hours. L refers to the number of items to be labeled with RFID tags per hour.

A refers to the cost of applying one RFID tag to an object. Depending on the particular implementation, these may be variable labor costs or fixed costs for robots.

H_T refers to hardware costs for one RFID tag. Because Eq. 5.4 applies to cases in which tags are not reused, these costs occur for each tagged item. Note that in Eq. 5.4 the overall costs for tags are directly proportional to the number of required tags. This must be adapted for pricing models with sales discounts.

CO refers to cost reductions that may be achieved due to cost-sharing models (compensations). If RFID tags remain on the products throughout several phases in the supply chain, costs related to the RFID application may be shared. In Eq. 5.4 this is modeled as compensation, which is proportional to the number of tagged items. However, other cost-sharing models may be considered.

5.5.2 Benefits

In this section we provide some guidance for assessing the monetary benefits of RFID introduction with respect to the use cases described in Sect. 5.3:

- Accelerating scan processes
- Extending scan processes for improving quality and efficiency
- Extending scan processes for narrowing recalls
- Reducing paper-based data management
- Automating asset tracking
- Reducing back-end interactions
- Unifying labels

Note again that not all benefits are monetarily quantifiable. For instance, performance gains due to better controlling may be hard to quantify in advance. Moreover, savings greatly depend on particularities of the application. However, we have identified some general patterns that can give some basic guidance in assessing RFID-related benefits. In the following sections we discuss each use case and outline equations that may serve as guides even though they will have to be adapted for each specific case.

5.5.2.1 Accelerating Scan Processes

One reason for applying RFID is to accelerate or completely automate the scanning of identifiers. We described this application in Sect. 5.3.1. In such cases, labor time can be reduced. The resulting benefits can be quantified as follows:

$$S_S = F_S \cdot T \cdot (S_A - S_R) \cdot P \qquad (5.5)$$

where

S_S	Savings due to the acceleration of scan processes
F_S	Frequency of scan transactions
T	Service life of the application
S_A	Time for scanning without RFID
S_R	Scanning time with RFID
P	Hourly cost of an employee

F_S denotes how many identifiers are scanned in the production processes within a certain amount of time. Multiplied by service life T, this results in the total number of scan transactions during the service life of the application. With term $(S_A - S_R)$ we can calculate the saved time per scan transaction. Here, S_A refers to the time needed for registering identifiers encoded with RFID alternatives such as barcodes or alphanumeric numbers. The time needed to scan an RFID tag is denoted by S_R. P denotes the labor costs and refers to the salary of the employees in charge of the scan transactions.

5.5.2.2 Extending Scan Processes for Improving Quality and Efficiency

RFID can help increase the visibility of production processes and help provide better insights into the activities on the plant floor. Extending process monitoring may be mandatory to meet regulations and customer requirements. Such applications are described in Sect. 5.3.2. The ability to better analyze the running production processes may also enable the detection of inefficiencies and thereby lead to productivity improvements. Furthermore, better quality control and more security in the process planning are among the potential benefits of extending process monitoring with RFID. However, it is notoriously difficult to determine the monetary benefit of extended process monitoring in advance. For instance, it may be difficult to foresee how many quality issues can be resolved and how much process efficiency can be increased if process monitoring is extended. Even the ex-post analysis of such measures is hard to accomplish.

Here, companies may have to operate with estimates, simulations, or test runs to forecast the potential for improvements. Possible gains may come from an increase in the plant's throughput due to process optimization and more reliable planning, less waste due to better quality controlling, smaller material buffers, smaller safety stocks due to less uncertainty in planning, and competitive advantages in compliance issues. How these factors can be quantified highly depends on the particular manufacturer. We therefore do not suggest an equation for this case.

5.5.2.3 Extending Scan Processes for Narrowing Recalls

As discussed in Sect. 5.3.3, narrowing the scope for recalls by using RFID data is a major driver for manufacturers to consider RFID adaption. Each recalled item may cause expenses for penalties and for conducting the recall. In addition, recalls of large quantities may negatively affect the manufacturer's reputation. Consequently, reducing the number of items in recalls is an issue of highest priority for many manufacturing companies.

Fine-grained RFID data collected in the production can help track down items that are potentially affected by production errors. In general, the smaller the batch size for tracking, the more that recalls can be narrowed. The size of this effect depends on the nature of errors and how well errors can be detected. For instance, errors may occur at single points in time or over a certain period. Errors in material processing may affect several items that were produced using these materials. We need to consider such particularities of the targeted production environment for calculating the scope of recalls. Equation 5.6 sketches the calculation of monetary benefits of reducing the batch size for tracking in a simple case. In the considered case, errors occur at a known single point in time and affect only a single item. In the equation we quantify the resulting benefits for extending scan processes for narrowing recalls:

$$S_R = F_R \cdot T \cdot (B_A - B_R) \cdot C_R \tag{5.6}$$

where

S_R	Savings due to narrowed recalls
F_R	Frequency of recalls
T	Service life of the application
B_A	Tracked batch size if RFID is not used
B_R	Tracked batch size if RFID is used
C_R	Costs related to the recall of one item

F_R denotes the frequency of errors that result in a recall of the potentially affected products. Multiplied by the service time T, this results in the total number of recalls during the service life of the application. The term $(B_A - B_R)$ determines how many items can be spared in a recall if the batch size for tracking is reduced. For instance, if tracking is done on the pallet level, all items in the affected pallet must be called back. RFID can help make tracking on the item level feasible, thereby narrowing the recall to the affected item. The product of the difference in the batch sizes and the total number of recalls equals the total number of items that could be spared in recalls. Multiplying this number with the cost C_R for recalling an item (such as penalties) determines the savings.

5.5.2.4 Reducing Paper-Based Data Management

As described in Sect. 5.3.4, improving data maintenance by RFID may reduce costs that result from errors in collected production data. This is because RFID can help automate data maintenance in some applications and thereby reduce human mistakes. Furthermore, RFID may accelerate or automate tasks of data maintenance and thereby save labor costs. We denote the basic factors that influence the savings due to improved data maintenance in the subsequent equation:

$$S_D = T \cdot (F_M \cdot C_M + F_F \cdot C_F + F_W \cdot C_W + F_E \cdot (T_A - T_R) \cdot P) \quad (5.7)$$

where

S_D	Savings due to easier data maintenance
T	Service life of the application
F_M	Frequency of data mix-ups that can be avoided by using RFID
C_M	Resulting costs of a data mix-up
F_F	Frequency that data entries are forgotten
C_F	Resulting costs of a forgotten data entry
F_W	Frequency that data entries are wrong
C_W	Costs resulting from a wrong data entry
F_E	Frequency of manual label scans
T_A	Needed time for making a data entry without RFID support
T_R	Needed time for making a data entry with RFID support
P	Hourly payment of an employee

In Eq. 5.7, we consider various types of errors in data maintenance. Depending on the particular production processes, data errors may result in expenses. For instance, incorrectly configured machines may produce waste, or forgotten bookings of finished steps may delay production. In the equation, the frequencies of mutual data mix-ups (F_M), forgotten entries (F_F), and wrong data entries (F_W) are multiplied by the costs resulting from the respective incidents and summed up. Another summand is the product of the frequency of data entries (F_E), labor cost in terms of salary per hours (P), and the time that can be saved if data entries are supported by RFID ($T_A - T_R$). For instance, such time savings can be achieved if machine configurations are read from RFID tags rather than being manually copied from accompanying documents. In the end, the resulting sum is multiplied with the planning period of the respective RFID application. This results in the overall savings that can be estimated due to improved data maintenance.

5.5.2.5 Automating Asset Tracking

In Sect. 5.3.5 we discussed the application of RFID for automatic asset tracking. Having the right assets available at the right time is crucial for seamless operation of a production plant. Missing assets can delay the production process and thereby result in financial losses. RFID can help implement automated tracking applications and thereby reduce expenses resulting from missing assets. Note that asset tracking can also be conducted without using RFID. However, depending on the application environment, barcodes or similar technologies may not be feasible. We included the factors influencing the monetary benefits of such an application in the following equation:

$$S_A = T \cdot (F_A - F_R) \cdot (OC + PE) \tag{5.8}$$

where

S_A	Savings due to automatic asset tracking
F_A	Frequency that assets are missing without RFID-based tracking
F_R	Frequency that assets are missing with RFID-based tracking
T	Service life of the application
OC	Opportunity costs resulting from downtimes of the production
PE	Penalties for delays resulting from downtimes of the production

In Eq. 5.8, the term $(F_A - F_R) \cdot T$ denotes how many cases of missing assets could be avoided in the respective planning period. For determining the total savings, this number is multiplied with the costs resulting from each incident of missing assets. These costs are reflected by the term ($OC + PE$). OC refers to the opportunity costs that result from unnecessary downtimes of a machine or production line. In particular, downtimes caused by missing assets are considered. Furthermore, downtimes may cause delayed deliveries from which penalties result. Expenses of this account are quantified in the variable PE.

5.5.2.6 Reducing Back-End Interactions

RFID allows storage of data with the corresponding object rather than in back-end databases. The business logic can thus be decoupled from the back-end system, and interactions with the back-end system can be reduced. This can help ensure fast access to production data without tuning the back-end databases and network infrastructure. We discussed in Sect. 5.3.6 how RFID may help reduce back-end interactions. However, deciding whether to store data in the back-end or on RFID depends on specific application constraints and on the available IT infrastructure. Thus, a financial trade-off on an abstract level is not feasible for this application.

However, applications that work on data from RFID tags are less vulnerable to system failures than centralized solutions are (no single point of failure). We estimate the monetary effect of such benefits with the subsequent equation:

$$S_B = T \cdot (F_B \cdot A_B \cdot T_B - F_R \cdot A_R \cdot T_R) \cdot (OC + PE) \tag{5.9}$$

where

S_B	Savings due to reduced back-end interactions
T	Service life of the application
F_B	Frequency of breakdowns in the back-end system
A_B	Number of products affected by a back-end failure
T_B	Duration of a breakdown in the back-end system
F_R	Frequency of failures in an RFID-based system
A_R	Number of products affected by a nonworking RFID tag
T_R	Time until a nonworking RFID tag is replaced
OC	Opportunity costs resulting from downtimes of the production
PE	Penalties for delays resulting from downtimes of the production

In Eq. 5.9, the term $F_B \cdot A_B \cdot T_B$ denotes how often the back-end system fails, how many production tasks would be affected thereby, and how long such failures last. For instance, in the case of a back-end failure, all production processes in a plant may stop if the operations are managed centrally. The term $F_R \cdot A_R \cdot T_R$ reflects the frequency of failures in an RFID-based system and the number of products that are affected. In general, malfunctions in RFID tags are more likely than failures of the back-end system. However, the effect of a failing tag will most certainly have a smaller scope than failures of the back-end system. With the term $(OC + PE)$, opportunity costs and costs for penalties that result from downtimes are considered. These are measured in cost per item per hour of delay time. Multiplied by the total service life of the application (T), this results in the total expenses for penalties and opportunity costs that can be saved by applying RFID.

5.5.2.7 Unifying Labels

Because different customers are likely to demand different labels for their shipments, label handling can be challenging for manufacturers. The ways that RFID may help meet this challenge were discussed in Sect. 5.3.7. One cost driver for printing labels is specialized multiformat printers. If barcode labels are replaced by RFID tags, RFID readers could be used for writing in multiple data formats. Note that readers may already be available at packing stations for other applications (such as in logistics); if so, no additional readers for label handling would be needed. However, this cost difference was already considered in Eq. 5.1 for general costs (in Sect. 5.5.1).

In some plants, labels are printed centrally to reduce the number of required printers. If RFID readers at the packing stations are used to write data on the labels, transporting the labels from centralized printing stations to the packing stations could be spared as well. Another cost factor related to label handling concerns the penalties for labels that customers cannot read. Here, RFID may be a more reliable solution in many cases. We estimate the monetary effect of such benefits by the following equation:

$$S_{\mathrm{L}} = T \cdot (((F_{\mathrm{B}} - F_{\mathrm{R}}) \cdot PE_{\mathrm{L}}) + C_{\mathrm{T}} \cdot L) \tag{5.10}$$

where

S_{L}	Savings due to unifying label handling with RFID
T	Service life of the application
F_{B}	Frequency that a barcode label is unreadable
F_{R}	Frequency that an RFID label is unreadable
PE_{L}	Penalties for an unreadable label
C_{T}	Costs for transporting a label from printer to packing station
L	Number of RFID labels applied per hour

In Eq. 5.10, the frequency that RFID tags are unreadable (F_{R}) is subtracted from the frequency that barcode labels are unreadable (F_{B}). Multiplied by the penalty per unreadable label (PE_{L}), this yields the penalties per hour that can be saved by switching from barcode to RFID. Further potential savings for this use case are calculated by multiplying the number of labels applied per hour with the cost for transporting a barcode label from the printer to a packing station ($C_{\mathrm{T}} \cdot L$). The cost could be spared if labels are written directly by RFID readers at the packing stations. The hourly savings due to reduced penalties and the savings due to decreased transportation costs for labels add up to the total savings per hour. The product of this sum and the application's total service life (T) equals the monetary benefit that could be achieved if RFID enables unifying labels for customers of the respective manufacturer.

5.6 Basic RFID Functionalities

In this section we describe functional building blocks that are common in existing RFID infrastructures. In particular, we elaborate on the following tasks:

- Filtering and enriching RFID data
- Storing RFID data
- Exchanging RFID data
- Detecting events in RFID data

Prominent middleware solutions provide modules that support these functionalities to a certain extent. Note that in this section, the software functionality is discussed independent of particular use cases; we address generic building blocks that are typically required in RFID infrastructures. Therefore, this section applies to RFID setups in logistics as well as in manufacturing and other applications. Special requirements for RFID systems in manufacturing are discussed in Sect. 5.7.

5.6.1 Filtering and Enriching RFID Data

Several operations must be applied to make use of RFID data: filtering of events, enrichment of the RFID data with process semantics and additional information, and inference of the data and reactions to events. Filtering of RFID data takes place on several levels, including during preprocessing or complex preprocessing (see Fig. 5.1). Read errors such as double reads are filtered on the lowest level. This is usually done by the device controller that provides the software interface to the readers. On a higher level, RFID data must be filtered with regard to the events of interest. At a gate, for instance, "appearance" events and "disappearance events" may be dropped and replaced by one aggregate "passing" event.

If additional information for an RFID tag is available, this information must be associated with the read event. This can be, for example, information in the user memory of a tag or corresponding sensor measures. On a higher level, the read events must be enriched with process semantics. That is, for each event it must be inferred which aspect of the production process is reflected.

To implement business logic based on RFID data, the event data must be further evaluated after it has been filtered and enriched. This is done by rules that define actions to be taken when certain observations have been made. Such rules can operate on an intermediate storage that holds RFID data collected during a certain period. Alternatively, the rules could run directly on the input stream of the read events. Any other system could be informed via application-level events, and actions could be triggered if certain predefined conditions occur.

5.6.2 Storing RFID Data

To decide on the overall architecture, it must be determined how and where to store the collected RFID data. This decision is affected by three major issues. First, one must be clear about where in the system the data should be evaluated. Second, it must be determined which degree of data aggregation is suitable for the intended use. Third, a policy about how to handle data in the long term should be devised and must specify how long the data are to be kept on a specific medium and whether data can be deleted after a certain time.

RFID-based information can be evaluated with two distinct objectives in mind. One is to control and monitor business processes. The other is to allow long-term analysis of the monitored activities and to document production. These two objectives lead to different storage requirements. Data for the control of business processes must be available very fast and usually comprises only recent information. For instance, automatic booking at the intake would require only the RFID data currently read and recent advanced shipping notices. Consequently, the required data should be kept in storage that may be relatively small but must be fast.

Long-term data analysis and business intelligence require different features. Typical tasks can be, for example, an investigation of how overall performance developed over time. Such tasks do not demand for real-time data. Instead, a database is needed to store the data in the long term. Data historians are tailored to store large amounts of time-stamped data. Using a data historian would allow the company to store information from every RFID readout, and such fine-grained information could be useful for detailed analyses of production processes. Mining tools could use the rich data source to search for patterns in the data and identify potential improvements in production.

Another method of data analysis is provided by data warehouses. Such software systems support data analysis by advanced tools for reporting and visualization (e.g., OLAP cubes). However, reports in data warehouses do not display raw data. Instead, the data are aggregated and evaluated to show key performance parameters. The raw read events must be mapped to the corresponding process step and enriched with context information in order to reflect the process semantics. Thus, data warehouses work on an excerpt of the complete data set.

In case of recalls, detailed information from the history is needed. In this case no aggregation is suitable. Fine-grained event data must be kept in a database (e.g., data historian) at least for the time a recall can occur. That is, the stored data must provide all information about the production process of parts that could be called back.

5.6.3 Sharing Information Along the Supply Chain

Fully exploiting the potential of RFID may involve exchanging captured RFID data with business partners. Increasing the transparency across organizational borders allows collaboration to be optimized. For this purpose EPCglobal is developing the EPC Network (EPCglobal 2005). The network will contain services for discovering and accessing information.

Several issues need careful consideration before RFID data can be exchanged with business partners. First, it must be decided which information should be available to whom. This requires management of roles and rights for all partners. Also, it is not sufficient to just provide the captured RFID data; the information must be aggregated and semantically enriched to be reasonably interpretable for the business partners.

Furthermore, the exchange format and the communication model must be determined. EPCglobal has proposed the physical markup language (PML) for exchanging RFID-related data (Floerkemeier et al. 2003). This XMLbased language allows RFID reads to be associated with additional data such as sensor measurements. It is used to encode the output of RFID readers in existing middleware solutions (Bornhövd et al. 2004). However, data exchange on the enterprise level may go beyond using pure PML or may be realized based on different standards.

With the specification of EPC Information Services (EPCIS), EPCglobal has released another standard relevant for exchange of RFID data (EPCglobal 2007). EPCIS leverages exchange of data that are related to Electronic Product Codes (EPCglobal 2006). The standard specifies data types as well as query interfaces and describes the use of EPCIS in a framework for exchanging RFID data. EPCIS supports interfaces for ad hoc queries and a callback interface for standing queries. It needs to be decided whether information should be pushed via callback functions to the recipient once it is available or whether data should be queried on demand via ad hoc requests. This trade-off can be decided based on the frequencies that data would be pushed or pulled.

Another important aspect is security. If trusted partners are getting insight into business operations, competitors may try to spy on the transferred information. It is therefore highly recommended to conduct a detailed security analysis before deciding on a concrete information-sharing arrangement.

5.6.4 Event Detection

It is necessary to evaluate the semantics of RFID read events in order to use the data for monitoring and controlling production processes. Evaluating these semantics is generally straightforward in logistics applications. For instance, the semantics of reading an object at the inbound gate is that the object has been taken in; an object that is read repeatedly by a smart shelf is stored in that shelf.

Processes on the plant floor can add complexity to the evaluation of the semantics. To be more precise: Observing an object at a certain point may not be sufficient to determine its status in the process. For example, registering a certain process step may involve detecting an object at the machine, registering a disappearance as it is loaded to the machine, checking that the machine is running, and observing the object for its appearance again. Evaluating such complex events requires explicit domain knowledge. That is, a model is needed of how a series of read events adds up to a complex event and how this is mapped to certain steps in the production process. Modeling and evaluating such domain knowledge should be supported by software systems for embedding RFID in manufacturing applications. This comprises a language for expressing evaluation rules and software components to execute these rules. Technologies for complex event processing (CEP) are suitable for processing such rules (Luckham 2001). A complex event rule comprises three parts: a pattern, a constraint, and an action part. The pattern part and the constraint part together describe the complex event that should be detected. The action part defines how the system must react when the defined event occurs. In that sense, CEP rules are similar to event condition action rules in active databases (Paton et al 1998). However, active databases focus on in-database events and do not support temporal constraints in RFID-specific queries well (Wang et al. 2006, Wu et al. 2006). Dedicated approaches for detecting complex events often use finite-state automata (e.g., Gehani et al. 1992, Coral8 2006) or Petri nets (e.g., Gatziu and Dirtrich 1994). Particularities of RFID-specific applications for CEP have been targeted in recent research (Wang et al. 2006, Wu et al. 2006, Gyllstrom et al. 2007). For RFID solutions in manufacturing, it must be determined which kind of events the software system should detect and which type of event detection technology should be implemented in the software system.

Besides the need for complex event detection, the streaming nature of RFID data must be considered. Read events float into the back-end system as continuous data streams. Readouts that belong to one complex event are usually distributed over time. Thus, modeling of processing rules requires operation on event streams. Research projects including STREAM (Arasu et al. 2003) and AURORA (Abadi et al. 2003) have targeted the issue of modeling and executing stream queries. Even though they lack expressiveness for complex events, the results may be adapted for RFID in manufacturing.

5.7 Requirements for the IT Infrastructure

Large-scale RFID applications are already in use, for instance in logistics. Such applications are supported by existing middleware solutions for capturing and processing RFID data. The typical use cases described above pose new requirements for the IT infrastructure. In this section we summarize technological issues that must be respected by software systems for a successful RFID integration in the manufacturing context. This comprises middleware issues that are important for any RFID integration in the business context. Additionally, we focus on the particular requirements of RFID applications on the plant floor. This covers aspects of integration with other systems, paradigms for data processing, and architectural and functional requirements. In particular, the following six issues for IT systems were derived; they all aim to support RFID integration on the plant floor:

- Providing RFID data to components of ISA-95 level 3
- Distributing business logic and data
- Supporting heterogeneous data sources
- Dealing with noise and uncertainty
- Supporting process analysis
- Supporting asset tracking

5.7.1 Providing RFID Data to Components of ISA-95 Level 3

In our case studies we observed similarities in the IT infrastructures that are used at the investigated plants. The used software systems supported a set of typical functionalities such as planning, controlling, and monitoring the production processes. Such general patterns and functional building blocks are reflected in the Purdue reference model for computer-integrated manufacturing (Williams 1992), which is the foundation of the ISA-95 standard (Brandl 2000, 2001, 2005). As already mentioned in Chap. 1, this standard gives a general reference model for the operations of any production plant. It divides the manufacturing domain into five levels of computing functionality, which are based on the hierarchy of the manufacturing enterprise (see Sect. 1.2.4). These levels are considered sufficient for the purpose of identifying necessary integration standards. The intention of the standard is to help reduce the costs and errors that arise when implementing interfaces between enterprise and production control systems. ISA-95 models enable companies to separate business processes from manufacturing processes. With the use of this standard, the integration of solutions from different suppliers promises to be less complex.

The lower three levels (0–2) in the reference model comprise the actual physical process as well as activities of sensing and machine control. Functions involved in monitoring and controlling the physical process are part of level 2 (for details, see Sect. 1.2.4). Components on the upper levels (3 and 4) are part of the back-end system. Components on level 3 serve functionality such as workflow control. Therefore, components on this level need information from level 2 components that monitor the production processes. In many use cases, these monitoring components will be based on RFID technology (Sects. 4.1–4.3, 4.5, 4.6). Thus, data from RFID readers must be fed into level 3 components in order to support the workflow management. Yet it is not feasible to directly link reader interfaces to components of level 3. Instead, RFID data must be preprocessed to extract information that is relevant for this system level. To be more precise, at least noise in the form of double reads or false disappearance events should be filtered. Furthermore, RFID reads should be aggregated to meaningful events, such as the completion of a certain production step.

Thus, interfaces to preprocessing components must be defined for level 3 components. These interfaces should support push-based communication to account for the event character of RFID readouts. Furthermore, a coding scheme for data exchange must be defined. This coding scheme should allow the association of identification numbers of RFID tags with related data and information about the read event.

5.7.2 Distributing Business Logic and Data

Distributing business logic and decoupling it from the back-end system was an explicit demand in some of the investigated enterprises (Sects. 4.1, 4.6). One reason for this demand is to improve reliability of the IT infrastructure by decentralization. For decentralized processes, the back-end system would no longer be a single point of failure, and overall reliability would be improved. For instance, production control and consistency checks should be conducted independently of the back-end system on terminal computers on the plant floor. In this setup, a failure of the back-end database or the communication network would have a smaller impact on production. Another reason for decentralizing business logic is to reduce the load on the back-end system and the communication network. Interviews with the IT staff of the visited companies disclosed that retrieving data from the back-end system often reveals bottlenecks in the system's performance. Processing and retrieving data locally at the terminal computers could reduce the network load and speed up the system.

To decentralize the business logic, the required input data must be distributed as well. For instance, consistency checks usually need data about the planned process. Additionally, information about the input materials is required. This can comprise static information about the input materials as well as historic data about past operations that were performed on them.

Required data from the production plan could be pushed to the terminal computers right after the plan is created. Therefore, the IT infrastructure must support identification of terminal computers that will be in charge of certain parts in a production plan, extracting the corresponding information from the production plan and pushing the data to the terminal computers' hard drives. Historic data about production steps could be stored directly at the manufactured product. This would require applying RFID tags with sufficient writable memory that move along with the product as it is processed.

From these observations we can derive two degrees of freedom concerning the distribution of data and logic. One concerns the question of where to store the data—on the tag or in the network? The other concerns where to place the business logic—near the production floor or on the edge?

Factors affecting the first degree of freedom, data on the tag versus data in the network, are visualized in Fig. 5.3. One factor captures the need for fast access to data. This is when the IT infrastructure must meet real-time requirements, and lookups in the back-end system are bottlenecks (e.g., see Sect. 4.1).

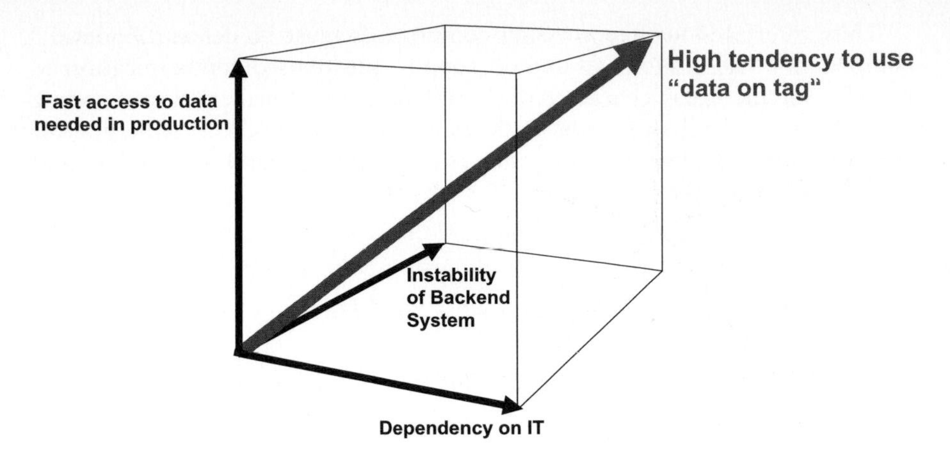

Fig. 5.3 Incentives for storing data on the tag

For such cases, data on tags may help ensure fast access to required information.

Another factor concerns the dependency of the production on the back-end system. High dependency means the production cannot run if the back-end system is down due to system failures. Storing production data on the tag can help establish emergency solutions that—at least temporarily—allow production to be kept up without having a connection to the back-end system (e.g., Sect. 4.6).

The third factor refers to the reliability of the back-end system. Storing data on the tag facilitates decentralization and helps avoid single points of failures. This can be relevant if the existing IT infrastructure is not optimized for reliability (for example, if no redundant systems are in place).

Figure 5.4 visualizes the second degree of freedom, which concerns processing the business logic locally versus at the edge. We identified five factors of influence, each reflected in a dimension of the pentagon in Fig. 5.4. Three of these factors are identical to the factors in the cube described above. This is because identified use cases for data on the tag often coincide with the deployment of business logic in the edge tier. In some cases, decentralization by means of data on the tag is beneficial only if the processing is decentralized as well.

The remaining two factors are high data volumes and processing/storing of aggregated data. These factors refer to issues of data preprocessing in terms of filtering and aggregation. Reducing the amount of data by means of filtering and aggregation is a necessity if data volumes are high. Pushing these operations to the edge tier can help avoid bottlenecks and improve the system's scalability. The higher the data volumes, the higher the incentive to push preprocessing to the edge tier.

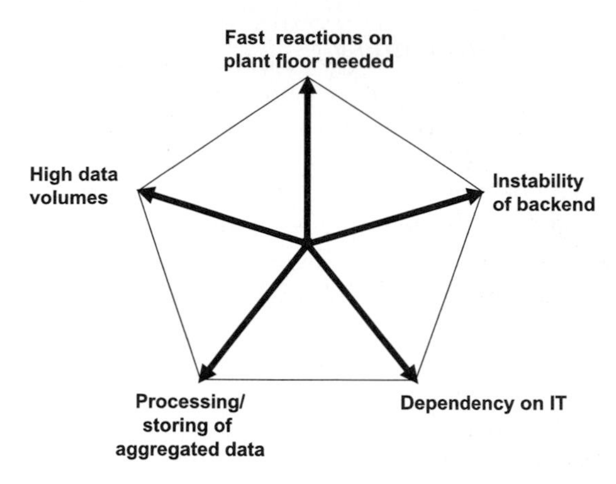

Fig. 5.4 Incentives for processing in the edge tier

Another reason for preprocessing and aggregation is that raw data are often not of interest. Instead, semantically enriched and meaningful information must be extracted from the stream of input data. Performing related operations in the edge layer avoids forwarding unnecessary information to the back-end system and reduces the overall system load.

Altogether, five factors support the tendency toward pushing processing to the edge tier. These are depicted in the pentagon in Fig. 5.4. Properties of a given production environment span a surface in this pentagon. The bigger this surface, the higher the motivation to push business logic toward the edge tier.

5.7.3 Supporting Heterogeneous Data Sources

One driver for adopting RFID in manufacturing is the demand for better insight into production processes (Sects. 4.2, 4.6). The demand for applications that help address quality issues was mentioned in several of the case studies (Sect. 4.6). Such applications need to keep track of each object in the production along with corresponding machine settings and sensor measurements of environmental conditions. Furthermore, control applications that evaluate production data in real time would have access to these information sources as well. For instance, the investigated manufacturer of cooling frames uses optical sensors to automatically route faulty products to a separate control station.

To enable such applications, RFID and sensor data must be correlated with information from different data sources. Therefore, a software module is needed that supports numerous ways of data acquisition and that provides an integrated view of the collected information. Besides RFID and barcode readers, relational databases, XML databases, and machines on the plant floor are relevant.

5.7.4 Dealing with Noise and Uncertainty

Using RFID and sensor data to monitor processes on the plant floor requires advanced evaluation of the raw input data. Sensor data are inherently overlaid with noise and distorted by measurement errors. Therefore, software for using sensor data in business processes must take the imperfect nature of the input data into account. This can be done by various filtering algorithms. A simple example would be to apply low-pass filtering over a set of input data or to build the average over multiple measurements in order to suppress noise.

Like sensor measurements, data from RFID readers also include errors. These errors are usually not corrupted readouts of RFID tags, since those errors are filtered out by checksums in the communication protocol in the reader. However, readers can miss RFID tags within their read range. This results in the false observation that the respective tag is absent (false negative).

5.7.5 Supporting Process Analysis

In several investigated plants, the predominant reason for considering RFID adoption was to be able to better narrow recalls (see Sect. 5.3.3). This is because RFID, compared with other auto-ID technologies, eases scan processes and makes data capture more feasible in more situations. Fine-grained data tracks would improve determination of faulty products. Because of the importance of narrowing recalls, software systems that support RFID applications in manufacturing must provide special analysis tools.

Another major potential of applying RFID at the plant floor is to gain more insight into the processes. The collected data may help detect inefficiencies and reasons for quality problems in production. However, analysis tools are needed to extract this knowledge from the data. These tools should be tailored to the application domain to extract performance measures in correspondence to the respective manufacturing processes as well as data from sensors and machines. For instance, identifying an unexpected quality problem due to humidity on the plant floor requires detecting a correlation between data from quality checks and sensor data taken during the processing of the respective products.

5.7.6 Supporting Asset Tracking

Using RFID for tracking crucial assets was a targeted application in several of the plants. This was desired to improve production schedule planning and to direct employees who fetch the assets. To meet this demand, tracking software must be integrated with RFID data and the planning application.

For integrating RFID data, the tracking application must be able to associate RFID readers or RFID position tags with spatial positions. The software must also be capable of inferring relocation of assets from the captured RFID read events. In some technical setups, this may be realized by simple rules that associate objects registered by certain readers with the respective positions. For instance, this can be suitable for assets in smart shelves that are equipped with RFID readers. However, inferring the position of assets may be more complex in other situations. For example, if assets are registered by mobile readers, asset tracking requires determining the reader's position first and then inferring the position of registered objects. Such particularities of the planned application setup pose requirements for the tracking software of choice. Here it is of special interest how object positions can be obtained and analyzed by the software system.

5.8 Hardware Issues on the Plant Floor

Plant floors of manufacturing companies are often hostile environments with extreme conditions (Sects. 4.1, 4.4, 4.6) Challenging factors include dirt, heat, the presence of metal, limited space, and others. RFID solutions for manufacturing must be able to cope with such conditions. Yet no standard solutions can serve the requirements of all production environments. Instead, an individual solution must be found for each case. In this section we discuss general hardware issues for RFID implementations in manufacturing. In particular, we focus on the following:

- Hostile physical conditions
- Presence of metal
- Demand for wireless communication
- Processes in close spatial proximity

5.8.1 Hostile Physical Conditions

In many of the studied companies, products are exposed to extreme conditions, most frequently heat (Sects. 4.1, 4.6). Special casings or foils can protect RFID tags from external influences. Protection against heat ensures slow heat conduction from the environment to the tag. The temperature at the tag can thereby be kept below its tolerance threshold if the exposure is only temporary. For Gen 2 tags, the maximum tolerance for heat is 85°C, and correct operation is ensured up to 65°C. Consequently, protection is needed for temperatures above 85°C. Some RFID tags are already shipped in protective cases; others can be wrapped in protective materials by converters that add materials to RFID inlays and produce complete labels.

5.8.2 Presence of Metal

Many metal objects were present on the plant floor of the investigated plants. Transportation units, machines, and even the products themselves can be made of metal. The presence of metal can influence communication with RFID tags because of signal attenuation, reflection, detuning, and eddy currents. How the communication is affected depends on the communication technology of the physical layer.

UHF tags communicate using a backscatter technology. HF and LF tags communicate by inductive coupling in the near field. In general, UHF tags suffer less from detuning effects and distortions due to eddy currents than LF and HF tags do. In contrast, near-field communicating tags face fewer problems caused by reflected waves. Such waves may totally cancel out a signal. Also, UHF signals are more actuated when passing materials. This holds especially for materials containing water. Yet all frequencies are shielded by metal.

Problems caused by detuning and eddy currents can be overcome by applying an isolation layer between the metal object and the tag. Reflections by metal objects in the near field can be blocked by radio frequency absorbers. Also, reflections may even be beneficial to direct signals around a metal object that would block the communication otherwise. Which effects have the most influence depends on particularities of the application environment. Thus, only field tests can fully clarify which hardware setup is most suitable.

5.8.3 Demand for Wireless Communication

Introducing new scan points and sensor devices on the plant floor requires the establishment of new communication channels. These can either be an extension of existing channels or the introduction of new communication means. Fixed wires may be sufficient for stationary scan points. However, wireless communication can become necessary if mobile hand readers are applied. Furthermore, the number of devices may render wired communication channels infeasible; this is especially the case if large numbers of sensor tags are installed on the plant floor. For easy deployment in the target setting, sensor tags are designed to work independently of fixed infrastructures and facilitate establishments of wireless ad hoc networks.

The wireless medium poses numerous challenges, which are already known from other application domains. Security problems, bandwidth changes, and energy issues of battery-powered devices are just some of them. Furthermore, the diversity on the physical layer must be considered in the system design. Devices with different operating frequencies and communication protocols must be integrated via gateways and hubs that bridge channels with different communication technologies.

5.8.4 Processes in Close Spatial Proximity

On some plant floors, different production steps take place in close physical proximity. In one of the investigated plants, different assembly steps are conducted within an one meter distance (see Sect. 4.1). Associating an RFID read event with the correct process step may be challenging if several steps are performed within the range of a reader. The read range for tags that communicate in the near field is generally easier to control than for backscatter tags. For tags that use inductive coupling, the read range is limited to the size of the near field. For instance, HF tags can typically be read out within about a 1-m distance. UHF tags can theoretically be read out from any distance, and praxis shows ranges up to about 7 meters. The relative long range and the possibility of reflections may cause UHF tags to be read from unexpected positions. In general, this makes UHF more difficult to handle than LF and HF if items in different process steps are physically close. However, directional antennas and limitations in the signal strength can help restrict readouts of UHF tags to the point of interest.

5.9 Current Motives and Open Potential for Using RFID

During our case studies we found that manufacturers had a range of motives for considering RFID applications. In this section we discuss these motives along with the taxonomy introduced in Sect. 1.3. That is, we categorize the incentives regarding strategic and operational use as well as those regarding inter-enterprise and intra-enterprise applications. Figure 5.5 shows how the cases that we studied in Chap. 4 fit into this taxonomy. For each case, the motives to use RFID determine the position in the taxonomy chart.

The manufacturer of airbags, AIR, aims to apply RFID for several reasons. One driver is the operational improvement of accelerating scan processes and thereby raising productivity. However, another argument is that AIR expects some of its costumers to demand RFID adoption in the near future, so the company wants to be prepared. At this point AIR can potentially gain strategic advantages by being ready for an RFID-enabled value chain. This could be a distinguishing factor and competitive advantage. Besides this, AIR is interested in using RFID for product labeling. Keeping RFID labels on the products would enable operations on the customers' side to be improved as well. This way AIR could provide additional services to its customers, which may account for further strategic advantages. Because the main driver for RFID at AIR is to increase productivity on the plant floor, this case belongs to the cluster of operational use within one enterprise. However, we also found aspects of strategic use for the enterprise as well as inter-enterprise operational use. Consequently, the case of AIR overlaps with these clusters in Fig. 5.5.

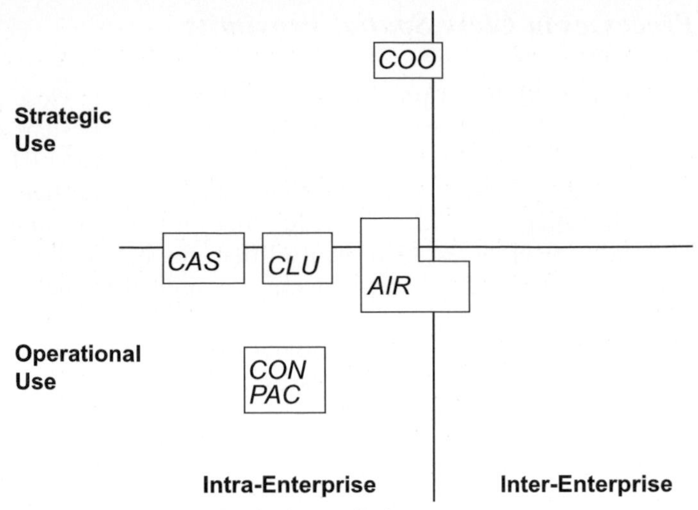

Fig. 5.5 Motives for using RFID in the studied cases

For the manufacturer of sliding clutches, CLU, the main driver of its RFID efforts is to improve product tracking to narrow recalls. A minor reason is to obtain more accurate reports about production status. Consequently, CLU's case is clearly motivated by operational intra-enterprise use. However, narrowed recalls may have the positive side effect of CLU's being better perceived by its customers. Therefore, this case includes some strategic aspects as well. Another effect is that efficient handling of recalls reduces the risk of disturbances in production at the customers' side. These side effects are reflected in the positioning of CLU in Fig. 5.5.

The manufacturer of packaging units, PAC, aims to use RFID to improve material tracking in the plant as well as to reduce dependency on its IT backend system. The targeted improvements are thus operational and intra-enterprise. Similarly, the motives of the manufacturer of connectors, CON, also mainly concern operational improvements within the company. In this case the main goal is to improve material tracking on the plant floor and to create more accurate production reports. Because only operational use within the enterprise is targeted in the cases of CON and PAC, these cases form a group in Fig. 5.5.

The motive for the RFID efforts at the manufacturer of cast parts, CAS, is to improve asset tracking within the plant. This application is mainly of local and operational use. However, reducing search times for assets would allow CAS to produce more flexibly and on shorter notice, which could lead to a strategic advantage (see Fig. 5.5).

The RFID application at the manufacturer of engine cooling modules, COO, is motivated solely by strategic issues. In this case no apparent operational benefits result from the use of RFID. Instead, RFID was introduced because of a customer's demand. Meeting this requirement was a strategic decision of COO. Additionally, the case has some aspects of inter-enterprise collaboration because COO provides the captured RFID data to the respective customers. However, this option is currently rarely used. In Fig. 5.5, COO's case is placed accordingly.

As depicted in Fig 5.5, the motives for RFID applications in the investigated cases are dominated by local operational improvements. Using RFID for strategic issues or across company boundaries is rarely targeted. Nevertheless, RFID applications targeting strategic issues and especially inter-enterprise application open up new potentials for the future. RFID may become a distinguishing factor by enabling additional services. An example is the provision of detailed production data to the customers.

The potential to use RFID for better strategic positioning may be extended further across whole value chains in use cases with inter-enterprise applications. Besides this, companies may collaborate on the operational level to use RFID in several steps of the value chain (Günther et al. 2006). This may enable collaborating partners to reduce costs and streamline processes. To fully exploit the potential of RFID, companies should extend the use cases from local and operational use toward applications involving inter-enterprise collaboration and strategic use. Collaborating on an operational level first and using RFID for company-specific strategies may be a viable path toward this goal.

5.10 Summary

In this chapter we presented insights on RFID in manufacturing that we gained during our case studies (see Chap. 4). Following a review of important terminology and a reference model for production processes, we presented the following common use cases for RFID in manufacturing:

- Accelerating scan processes
- Extending scan processes for quality and efficiency
- Extending scan processes for narrowing recalls
- Reducing paper-based data management
- Automating asset tracking
- Reducing back-end interactions
- Unifying labels

These seven use cases were found repeatedly during the conducted case studies. We then compared possible realizations with barcode and RFID technology against the background of these use cases. This was followed by a cost and benefits discussion for using RFID in these listed cases.

Furthermore, we dealt with implications of the IT infrastructure that we derived from our investigations. We discussed basic functionalities of RFID applications as well as special requirements and challenges for IT infrastructures that support RFID in manufacturing. The distribution of data and logic in the system was emphasized. Beyond this, we highlighted challenges on the hardware level that are particularly important to the manufacturing domain when applying RFID technology.

The chapter concluded with a discussion of current motives and open potentials for using RFID. Here, we categorized use cases regarding strategic versus operational use as well as inter-enterprise versus intra-enterprise use. Based on this, we outlined open potentials and possible future developments for RFID.

Chapter 6
Conclusions and Outlook

In this book we gave an overview of RFID in the manufacturing industry, based on six case studies involving small and midsize manufacturing companies. All companies are headquartered in Germany, but most of them have considerable export activities. Their sizes range between several hundred and 18,000 employees, and they typically occupy central positions in complex supply chains, acting as first- or second-tier suppliers to large OEMs, especially in the automotive industry. They typically have an advanced IT infrastructure organized around an ERP system and an MES.

As is true for any case-based analysis, we cannot claim that our insights are representative or have general validity. Nevertheless, we believe the lessons we learned may be helpful for many decision makers independent of geography, industry, and size. Our belief is based on the considerable overlap of the insights we obtained from the companies in our sample, both concerning their current situation and their plans and prognoses for the future.

All of the companies we surveyed see considerable potential for RFID in manufacturing, both in terms of efficiency and effectiveness. Efficiency potential was seen by possible speed-ups in production, lower error rates, reduced shrinkage, improved asset tracking, and less downtime—all contributing to a higher overall productivity of manufacturing operations. Effectiveness potential was seen concerning a variety of additional services that manufacturing companies could provide their customers. The tagging of objects itself often leads to immediate productivity gains at OEMs because they can use the tag and the additional information in their own operations. Moreover, the tracing of faulty components back to their manufacturing history is of great importance in security-sensitive industries such as the automotive sector.

For these positive potentials to come true, it is crucial that RFID not form a technology island but be tightly integrated into existing IT infrastructures. Although RFID is a technology with important implications for many manufacturing processes, the mapping of these processes to corresponding IT structures remains the task of MES and ERP systems. Such systems need to be

O. Günther, W. Kletti, U. Kubach, *RFID in Manufacturing*
DOI: 10.1007/978-3-540-76454-0, © Springer 2008

adapted to take advantage of the rich data that are becoming available through RFID. Appropriate filtering techniques need to be put in place to make sure that ERP and MES components receive the relevant information in the appropriate granularity—an information logistics problem that many ERP and MES companies are actively addressing. Moreover, companies must carefully consider how to distribute storage and processing in the resulting multitier IT architecture that ranges from RFID tags and sensors to data warehouses and business intelligence tools—the modern version of the information pyramid pictured in Fig. 1.1.

Given this positive outlook, one must ask why RFID in manufacturing is not more widespread than it is today. We see three major reasons:

First, the technology is not always sufficiently robust and reliable enough to be used in a manufacturing environment. Every manufacturing environment is different, and the validity of related concerns varies accordingly. The kinds of materials being processed especially play a major role in this context. According to our observations—not only regarding the case studies but also in other companies we surveyed—these technical roadblocks are gradually becoming smaller. New RFID technologies are often more robust with respect to adverse environmental conditions, and they operate with very low error rates. This makes them attractive to a much broader range of manufacturing companies than was the case only a few years ago.

Second, in many cases barcodes are an attractive and economic alternative to RFID. As discussed in the previous chapters, barcodes do not offer the same functionality as RFID, and they also expose different sensitivities to adverse environmental conditions (dirt, heat, and so on). But we have described several scenarios in which barcode solutions fulfilled the requirements in a satisfactory manner, and they did so at a much lower price than RFID. In other words, the advantages that can potentially be achieved may not be worth the price.

Third, and maybe most important, we have seen numerous situations in which the costs and benefits of RFID do not occur at the same enterprise, a situation that leads to investment deadlocks. Many supply chains find themselves in a kind of prisoner's dilemma: Even though the introduction of RFID would improve the productivity of the supply chain as a whole, none of the participating companies goes ahead with it. The reason is that everybody is afraid of being stuck with the cost of the introduction while the benefits occur elsewhere. This dilemma is particularly visible at many first- and second-tier suppliers; they are afraid of paying for RFID introduction while most of the productivity gains benefit the OEMs downstream the supply chain. At this point, most working RFID applications are therefore motivated by short-term operational improvements for which a positive return on investment (ROI) over a few years seems likely. Without such operational improvements in sight, the adoption decision is a much harder one because it becomes purely strategic. The question then is whether there is an early-mover advantage for suppliers: Is it advantageous to invest in RFID even though the short-term ROI is negative? This depends on whether the investment may help the supplier achieve a better competitive position in the long term. That is, it depends on the future behavior of the OEM, and in particular on its intentions to favor suppliers that use RFID in the future.

In our view, it is indeed the OEM that is most likely able to break the deadlock. OEMs that design and communicate an RFID strategy early on will be able to work with their suppliers on joint initiatives, which is likely to lead to competitive advantages. As shown by the Metro Group in retail, it can be advantageous to be an early mover (Günther and Spiekermann 2005). In many cases, however, a successful RFID introduction will imply some cost-sharing arrangement among the supply chain partners. This may involve cross-payments or nonmonetary compensation such as data-sharing agreements.

If a sufficient number of OEMs are willing to break the existing logjam, we will see RFID moving from a technology that is used mostly operationally and within enterprises (see Fig. 1.9) to being one that fulfills strategic purposes and helps improve the efficiency and effectiveness of inter-enterprise collaboration and supply chains as a whole.

References

Abadi DJ, Carney D, Çetintemel U, Cherniack M, Convey C, Erwin C, Galvez EF, Hatoun M, Maskey A, Rasin A, Singer A, Stonebraker M, Tatbul N, Xing Y, Yan R, Zdonik SB (2003) Aurora: a data stream management system. In: Proceedings of the SIGMOD conference, San Diego, 9–12 Jun 2003, p 666

Arasu A, Babcock B, Babu S, Datar M, Ito K, Motwani R, Nishizawa I, Srivastava U, Thomas D, Varma R, Widom J (2003) STREAM: the Stanford Stream Data Manager. IEEE Data Eng Bull 26(1):19–26

Atkinson RD, McKay AS (2007) Digital prosperity: understanding the economic benefits of the information technology revolution. The Information Technology & Innovation Foundation., Washington, DC. http://www.itif.org. Accessed 8 Aug 2007

Bornhövd C, Lin T, Haller S, Schaper J (2004) Integrating automatic data acquisition with business processes—experiences with SAP's auto-ID infrastructure. Paper presented at the 30th international conference on very large databases, Toronto, 29 Aug–3 Sept 2004

Brandl D (ed) (2000) ANSI/ISA-95.00.01 Enterprise–control system integration. Part 1: models and terminology. ISA, Research Triangle Park, NC

Brandl D (ed) (2001) ANSI/ISA-95.00.02 Enterprise–control system integration. Part 2: object model attributes. ISA, Research Triangle Park, NC

Brandl D (ed) (2005) ANSI/ISA-95.00.03 Enterprise–control system integration. Part 3: activity models of manufacturing operations management. ISA, Research Triangle Park, NC

Chappell G, Ginsburg L, Schmidt P, Smith J, Tobolski J (2003) Auto-ID on the Line: The Value of Auto-ID Technology in Manufacturing. White Paper ACN-AUTOID-BC005. Auto-ID Centre. http://www.autoidlabs.org/single-view/dir/article/6/123/page.html (Date: July 18, 2007)

Coral8, Inc (2006) Complex Event Processing: Ten Design Patterns. White Paper. http://complexevents.com/wp-content/uploads/2007/04/Coral8DesignPatterns.pdf (Date: July 18, 2007

DeJong CA (1998) Material handling tunes in. Automotive Manufacturing & Production, Vol. 110(7): 66-9

Deshpande AD, Singh AK (2006) Tagging the supply chain. SAP INFO 141:70

EPCglobal (2005) Standards. http://www.epcglobalus.org/StandardsDevelopment/EPCglobalStandards/tabid/185/Default.aspx. Accessed 18 Jul 2007

EPCglobal (2006) EPCglobal tag data standards version 1.3, ratified specification, 8 Mar 2006. http://www.epcglobalinc.org/standards/tds/tds_1_3-standard-20060308.pdf

EPCglobal (2007) EPC Information Services (EPCIS) version 1.0 specification, ratified standard, 12 Apr 2007. http://www.epcglobalinc.org/standards/epcis/epcis_1_0-standard-20070412.pdf

Floerkemeier C, Anarkat D, Osinski T, Harrison M (2003) PML core specification 1.0. Available via Auto-ID Center. http://citeseer.ist.psu.edu/floerkemeier03pml.html

Franklin (ed) (2005) Business 2010: embracing the challenge of change. Report from the Economist Intelligence Unit. http://www.eiu.com

García A, McFarlane D, Fletcher M, Thorne A (2003) Auto-ID in Materials Handling. White Paper CAM-AUTOID-WH013. Auto-ID Centre. www.autoidlabs.org (Date: July 18, 2007)

Gatziu S, Dirtrich K (1994) Detecting composite events in active database systems using Petri nets. In: Widom J, Chakravarthy S (eds) Proceedings of the 4th international workshop on research issues in data engineering: active database systems, Houston, 14–15 Feb 1994, pp 2–9

Gehani N, Jagadish H, Shmueli O (1992) Event specification in an active object-oriented database. In: Proceedings of the ACM SIGMOD international conference on management of data, San Diego, 2–5 Jun 1992, pp 81–90

GS1 (2007) GS1 homepage. http://www.gs1.org

Günther O, Ivantysynova L, Teltzrow M, Ziekow H (2006) Kooperation in RFID-gestützten Wertschöpfungsnetzen. Industrie Management 22:41–44

Günther O, Spiekermann S (2005) RFID and the perception of control: the consumer's view. Commun ACM 48(9):73–76

Gyllstrom D, Wu E, Chae H, Diao Y, Stahlberg P, Anderson G (2007) SASE: complex event processing over streams. In: Proceedings of the 3rd biennial conference on innovative data systems research, Asilomar, CA, 7–10 Jan 2007

Hackenbroich G, Bornhövd C, Haller S, Schaper J (2005) Optimizing business processes by automatic data acquisition: RFID technology and beyond. In: Roussos G (ed) Ubiquitous & pervasive commerce. Springer, Berlin, pp 33–52

Ivantysynova L, Ziekow H, RFID in der Produktion: Eine Fallstudie aus der Airbagindustrie, GI Workshop RFID-Einsatz in kleinen und mittelständischen Unternehmen (2007) .

Kletti J (2006) MES—manufacturing execution system. Moderne Informationstechnologie zur Prozessfähigkeit der Wertschöpfung. Springer, Berlin

Kletti J (2007a) Manufacturing execution system—MES. Springer, Berlin

Kletti J (2007b) Konzeption und Einführung von MES-Systemen Zielorientierte Einführungsstrategie mit Wirtschaftlichkeitsbetrachtungen, Fallbeispielen und Checklisten. Springer, Berlin

Lampe M, Strassner M (2003) The potential of RFID for moveable asset management. Workshop on ubiquitous commerce at Ubicomp 2003, Seattle, October 2003

Luckham D (2001) The power of events: an introduction to complex event processing in distributed enterprise systems. Addison-Wesley Longman, Boston

Niederman F, Mathieu R, Morley R, Kwon I (2007) Examining RFID applications in supply chain management. Commun ACM 50(7):92–101

OPC Foundation (2007) OPC. http://www.opcfoundation.org. Accessed 18 Jul 2007

Open Applications Group (2007) Open Applications Group Integration Specification (OAGIS). http://www.openapplications.org. Accessed 18 Jul 2007

Paton NW, Schneider F, Gries D (eds) (1998) Active rules in database systems, 1st edn. Springer, New York

RosettaNet (2007) RosettaNet standards. http://www.rosettanet.org/RosettaNet. Accessed 18 Jul 2007

Schultz C (2007) A roadmap to RFID integration on an SAP-centric platform. SAP NetWeaver Mag

Spiess P (2005) Collaborative business items: decomposing business process services for execution of business logic on the item. Paper presented at the European workshop on wireless sensor networks, Istanbul, 31 Jan–2 Feb 2005

Tellkamp C, Quiede U (2005) Einsatz von RFID in der Bekleidungsindustrie—Ergebnisse eines Pilotprojekts von Kaufhof und Gerry Weber. In: Fleisch E, Mattern F (eds) Das Internet der Dinge. Springer, Berlin, pp 143–160

Wang F, Liu S, Liu P, Bai Y (2006) Bridging physical and virtual worlds: complex event processing for RFID data streams. Advances in database technology—EDBT 2006. 10th international conference on extending database technology, Munich, 26–31 Mar 2006. Lecture notes in computer science, vol 3896. Springer, Berlin, pp 588–607

Williams T (ed) (1992) The Purdue Enterprise Reference Architecture—a technical guide for CIM planning and implementation. ISA, Research Triangle Park, NC

Wolf K, Holm C (1998) Total Cost of Ownership: Kennzahl oder Konzept? Information Management, 13(2), 19–23

Wu E, Diao Y, Rizvi S (2006) High-performance complex event processing over streams. In: Proceedings of the 2006 ACM SIGMOD international conference on management of data, Chicago, 2006, pp 407–418

Subject Index

Printing: Krips bv, Meppel, The Netherlands
Binding: Stürtz, Würzburg, Germany